OPERATION & MAINTENANCE
OF
SEWAGE TREATMENT PLANTS

by   William Cameron & Frank L. Cross, Jr.

a **TECHNOMIC**® publication
TECHNOMIC Publishing Co., Inc.
265 Post Road West, Westport, CT. 06880

**OPERATION & MAINTENANCE
OF
SEWAGE TREATMENT PLANTS**

©TECHNOMIC Publishing Co., Inc. 1976
265 Post Road West, Westport, CT 06880

Printed in U.S.A.

Library of Congress No. 76-19504

ISBN 0-87762-212-4

TABLE OF CONTENTS

ACKNOWLEDGMENTS

The authors would like to express gratitude to all who have contributed to this text.

Special acknowledgment is given to Ralph Metcalf for many of the photographs used.

Acknowledgment is given to EPA for the use of excerpts from many of their wastewater treatment publications.

We would like to thank Richard A. Young, Editor of POLLUTION ENGINEERING Magazine for allowing us to reprint several articles that previously appeared in the magazine.

PREFACE

This is a basic text concerning the operation and maintenance of municipal waste treatment facilities. The manager and the operator alike should be able to use this text to focus on typical problem areas and affect a needed operational change or instigate a maintenance program.

The control of water pollution will require numerous types of efforts and costs. The efficiency of a waste treatment facility and the cost of control is directly related to the efforts of the operator and the plant management.

The benefits of efficient water pollution control will be both tangible and intangible. The tangible benefits include "damage reductions" (health, esthetics, etc.) Intangible benefits involving sight, color, odor, taste and other esthetic qualities of clean water are expected.

The operation of the plant will have a direct bearing on these benefits and whether or not they could be considered cost effective.

Please note the following chart:

Total capital, operating and maintenance costs required to meet current standards in 1968 and 1976 (1967 dollars expressed in millions) (1)

	1968 Required capital	1968 Operating and maintenance	1968 Total annualized costs*/	1975 Required capital	1976 Operating and **/ maintenance	1976 Total annualized costs*/
Municipal	16,810	731		19,382	843	
Industrial	8,966	1,320		10,752	1,582	
Total	25,776	2,051	3,880	30,134	2,425	4,562

*/ All capital costs were annualized on the basis of 25 years and 5 percent interest.

**/ The 1976 operating and maintenance figures were derived by applying the proportion of O&M costs to total capital in 1968 to the total 1976 capital.

(1) State-of-Art Review: Water Pollution Control Benefits & Costs -- Vol. 1, EPA-600/5-73-008a, USEPA, Office of Research & Development, Oct. 1973.

CHAPTER 1

INTRODUCTION

The operation and maintenance of a wastewater treatment plant should not be taken lightly. Municipal sewerage facilities are extremely expensive and require competant operators and managers to insure that this investment is properly protected.

As the *operator* is the key to the operation and maintenance aspects of a waste treatment plant, is is important that trained personnel be employed for the various positions. The general skills required for operators has been suggested by EPA. [1]

"The skill requirements outlined below are minimal for successful performance of specific required duties. These are only a guide; additional requirements for the particular plant location should be checked.

- *Supervisory Personnel* (level of ability depends on size and type of plant) — high school education or equivalent, should display better than average ability to:
  1. Use and manipulate basic arithmetic and geometry.
  2. Think in terms of general chemistry and physical sciences.
  3. Understand biological and biochemical actions.
  4. Grasp meaning of written communications.
  5. Express thoughts clearly and effectively, both verbally and in writing.

  In addition, supervisory personnel are often responsible for:
1. Public relations
2. Bookkeeping
3. Analysis and presentation of data
4. Budget requests
5. Report writing
6. Personnel
7. Safety educational program
8. Contracts, specifications and codes
9. Estimates and costs
10. .Plant library
- *Laboratory Technicians* — require training in laboratory procedures and mathematics
- *Operating Personnel* — require training in:
  1. Fundamentals of wastewater treatment processes, including chemistry and biology.
- *Maintenance Personnel* — must be familiar with and capable of:

1. Mechanical repairs
2. Electrical and electronic repairs."

Generally, the manpower requirement for plant operation is not clear cut, but some guidelines have been developed by EPA as an indicator. The following table [1] "shows ranges of the number of personnel which would be required to operate various modes of treatment systems. Each plant may have its own particular operating mode, depending on the number of components which make up liquid and sludge treatment, along with administrative and general plant functions. The advanced waste treatment plants were not included in this table because of the lack of reliable manpower estimates for this classification."

Thus anywhere from four to 80 people may be required to handle the various types of larger sized waste treatment plants. It is obvious that a detailed management system is needed to properly operate these facilities.

For the smaller facilities, the State of Florida suggests the following manpower requirements:

17—16.13 Operation of Domestic Wastewater Facilities.

(1) All owners of domestic wastewater facilities shall employ on-site at least a Class "C" certified operator as specified below:

Facility Classification as determined by 17—16.12 (2).	Minimum on-site time required of a certified operator
Class A	24 hrs/day, 7 days/week
Class B	16 hrs/day, 7 days/week
Class C 100,000 gpd and over	8 hrs/day, 7 days/week
Class C under 100,000 gpd	An average of one hour/day
Class C under 25,000 gpd	1 inspection/day, 7 days/wk for minimum total of 2 hrs.

The requirements for the licensing of operators is based upon education and experience. In most areas, some type of license is required by the state for the operator of a municipal waste treatment facility. The State of Florida's requirements are as follows: [2].

Each of the unit operations in a sewage treatment plant performs a physical, chemical, or biological function.

The personnel at a wastewater treatment facility must be familiar with the role of each of these operations as it effects plant efficiency as well as how to correct operational problems and to maintain this equipment on a routine basis.

There is no substitute for on-the-job training, but this text has been developed to acquaint the reader with the fundamentals of wastewater treatment plant operation and maintenance.

PLANT MANPOWER REQUIREMENTS*

Type of Plant	Average Capacity (MGD)									
	1	3	5	10	20	35	50	65	80	100
Primary	4.5–6	6.5–7.5	7.5–9	10–13	15.5–19	22–27	29–34	34–41	40–49	50–59
Secondary (including Trickling Filter)	6–7	7.5–9.5	9.5–11.5	13–16	19.5–24.	28–34	37–44	45–53	53–61	63.5–76.5
Secondary (including Activated Sludge)	7–8	9.5–10.5	11.5–13	15–18	23–26	33–38	43–49	51–59	61–69	71–82

*Based on a preliminar study performed by Black and Veatch for the EPA.

	DESCRIPTION	DESIGN FLOW		
		Class A	Class B	Class C
Level I	Chemical and/or physical processes providing a high degree of treatment including advanced treatment other than polishing ponds.	3.0 MGD or greater	0.5 MGD up to 3.0 MGD	2,000 gpd up to 0.5 MGD
Level II	Activated sludge process or modification, other than extended aeration.	5.0 MGD or greater	1.0 MGD up to 5 MGD	2,000 gpd up to 1.0 MGD
Level III	Extended aeration process	8.0 MGD or greater	2 MGD up to 8 MGD	2,000 gpd up to 2 MGD
Level IV	Trickling Filter process	10 MGD or greater	3 MGD up to 10 MGD	2,000 gpd up to 3 MGD

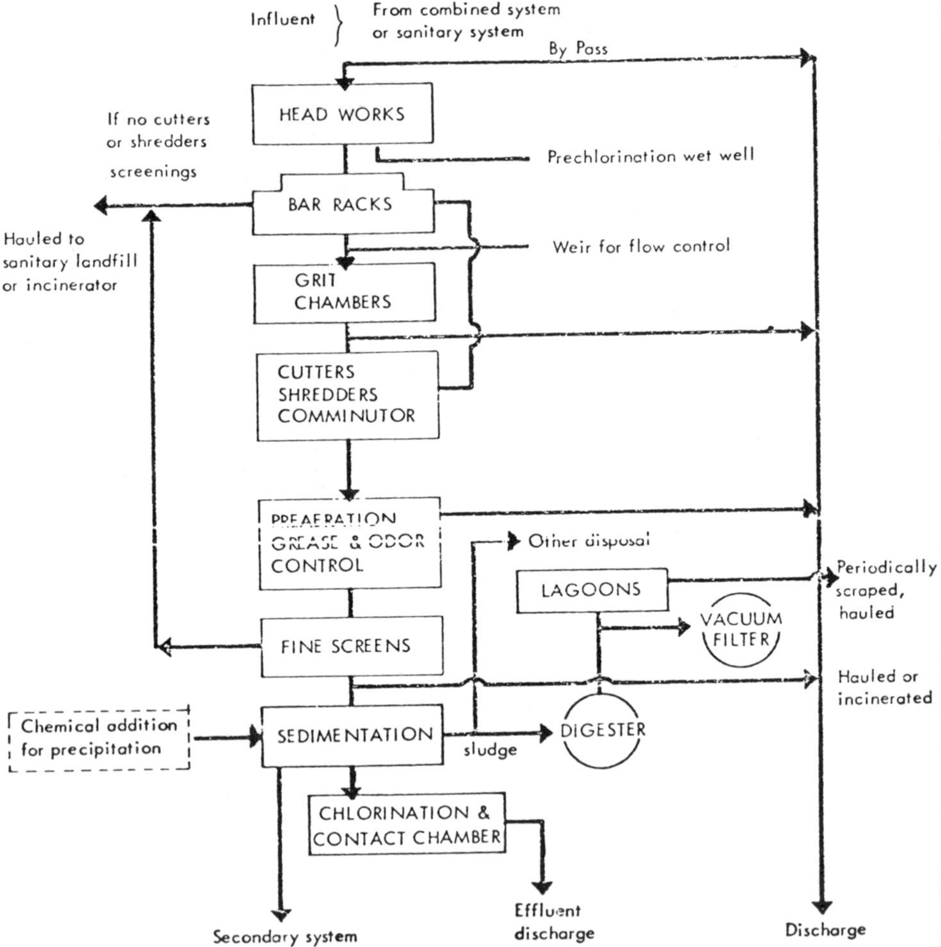

Figure 1. Unit Operations in the Waste Treatment System.

With EPA and state regulatory agencies making inspections, the municipality is constantly having attention focused on its operation.

[1] "The evaluation of a wastewater treatment plant consists of an in-depth analysis of the following basic elements:

- Plant performance
- Operational problems
- Operating personnel
- Sampling and testing program
- Laboratory facilities
- Maintenance data program.

"Information and data for each element are gathered and analyzed in four interrelated phases, namely:

1. Preparation for site visit
2. On-site inspection
3. Problem identification
4. Total plant evaluation.

"The estimated evaluation time will depend on the size and complexity of the plant, the amount of preparation on the part of the investigator(s), and the willingness of the plant personnel to cooperate with the investigator(s)."

The maintenance of a plant depends upon an adequate inventory of parts and a schedule of preventive maintenance.

[1] "It is imperative that a record be kept of the service requirements of every piece of major equipment in the plant and when and how frequently service is required. Therefore, a system is needed to keep a complete record of all maintenance requirements. Such a system should provide a permanent record of all maintenance work together with the advanced scheduling of preventive maintenance for an entire year. The system should also provide the maintenance work schedule for any given day. To be efficient, the system should contain the following five files:

1. preventive maintenance records
2. the preventive maintenance schedule for each piece of equipment
3. specifications on each major piece of equipment, the supplier, and where spare parts can be purchased
4. spare parts inventory, and
5. instructions for operation and maintenance of each item of major equipment.

Thus this text should provide techniques for the operation and maintenance of a waste water treatment facility for both the operator and the manager.

BIBLIOGRAPHY

1. *Procedures for Evaluating Performance of Wastewater Treatment Plants,* Contract No. 68-01-0107, USEPA, Office of Water Programs.
2. Rules of the Department of Pollution Control, State of Florida, Chapter 17–16 (Suppl. No. 49).

CHAPTER 2

TYPES OF WASTE TREATMENT PROCESSES

INTRODUCTION

Upon reaching a wastewater treatment plant, the wastewater flows through a series of treatment processes which remove the wastes from the water and reduce its threat to the public health before it is discharged from the plant. The number of treatment processes and the degree of treatment usually depend of the uses of the receiving waters. Treated wastewaters discharged into a small stream used for a domestic water supply and swimming will require considerably more treatment than wastewater discharged into water used solely for navigation.

Next, the wastewater will generally receive primary treatment. During primary treatment some of the solid matter carried by the wastewater will settle out or float to the water surface where it can be separated from the wastewater being treated.

Secondary treatment processes usually follow primary treatment and commonly consist of biological processes. This means that organisms living in the controlled environment of the process are used to partially stabilize (oxidize) organic matter not removed by previous treatment processes and to convert it into a form which is easier to remove from the wastewater.

Waste material removed by the treatment processes goes to solids handling facilities and then to ultimate disposal.

Waste treatment ponds may be used after pretreatment, primary treatment, or secondary treatment. Ponds are frequently constructed in rural areas where there is sufficient available land.

PRETREATMENT

The purpose of preliminary treatment is to remove from the sewage some of its constituents which can clog or damage pumps, or interfere with subsequent treatment processes. Preliminary treatment devices are, therefore, designed to:

1. Remove or to reduce in size the large suspended or floating organic solids. These solids consist of pieces of wood, cloth, paper, garbage, together with some fecal matter.
2. Remove heavy inorganic solids such as sand, gravel and possible metallic objects all of which are called grit.
3. Remove excessibe amounts of oils and greases.

A number of devices or types of equipment are used to obtain these objectives.

Although flow measuring devices are not for *treating* wastes, it is necessary to know the quantity of wastewater flow so adjustments can be made on pumping

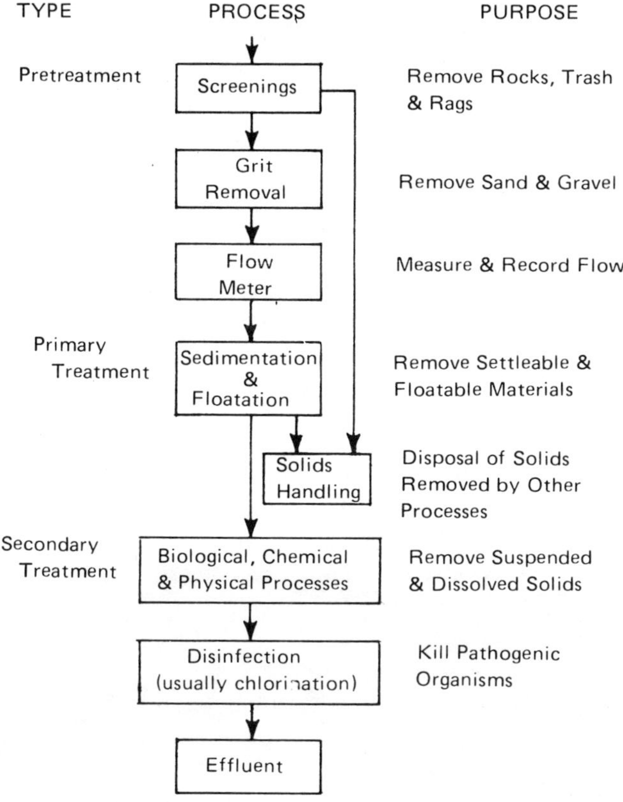

Figure 1. Treatment Processes.

rates, chlorination rates, aeration rates, and other processes in the plant. Flow rates must be known, also, for calculations of loadings on treatment processes and treatment efficiency. Most operators prefer to have a measuring device at the headworks of their treatment plants.

The most common measuring device is a Parshall Flume. Basically it is a narrow place in an open channel which allows the quantity of flow to be determined by measuring the depth of flow. It is a widely used method for measuring wastewater because its smooth constriction does not offer any protruding sharp edges or areas where wastewater particles may catch or collect behind the metering device.

Another measuring device used in open channels is a weir. A weir is a wall placed across the channel over which the waste may fall. It is usually made of thin metal and may have either a rectangular or V-notch opening. Flow over the weir is determined by the depth of waste going through the opening. A disadvantage of a

Table 1. Pretreatment Operation & Maintenance Parameters.

PROCESS	OPERATION				MAINTENANCE		
	Cleaning	Odor Control	By-Pass	Solids Disposal	Lubrication	Calibration	Annually
Screening							
Mechanical	Automatic	Hose Down	Manual Screen	2.	Weekly	—	4.
Manual	Daily (at least	Hose Down	Parallel screen	2.	—	—	—
Shredding							
Barminuter	Automatic	Hose Down	Screen	—	3.	—	4.
Comminuter	Automatic	Hose Down	Screen	—	3.	—	4.
Cutting Pump	Automatic	—	Screen	—	3.	—	4.
Grit Removal						—	
Chamber		Hose Down	Parallel	Burial	—	—	—
Air Degritter	1.	Prechlorination	—	Burial	—	—	—
Flow Measurement							
Weirs	1.	Prechlorination	—	—	—	—	Annually
Parshall Flume	1.	Prechlorination	—	—	—	Annually	—
Venturi Meter	1.	Prechlorination	—	—	—	Annually	—
Magnetic Flow Meter	1.	Prechlorination	—	—	—	Monthly	—

Notes:

1. Manually as required.
2. Burial or incineration.
3. As recommended by manufacturer.
4. As recommended by manufacturer or more often as determined by experience.

weir is the relatively dead water space that occurs just upstream of the weir. If the weir is used at the head end of the plant, organic solids may settle out in this area. When this occurs, odor and unsightliness can result. Also, as the solids accumulate the flow reading may become incorrect.

A good measuring device for flows of treated or untreated wastewater is a Venturi meter. It is a special section of contracting pipe, and it measures flow in much the same way as a Parshall Flume. It does not offer any sharp obstructions for particles to catch on. Magnetic flow meters also are being used successfully to measure wastewater flows.

Primary Treatment

We have previously discussed the reduction in velocity of the incoming waste to approximately one foot per second in order to settle out heavy inorganic material or grit. The next step in the treatment process is normally called *sedimentation* or *primary treatment*. In this process the waste is directed into and through a large tank or basin. Flow velocity in these tanks is reduced to about 0.03 foot per second, allowing the settleable solids to fall to the bottom of the tank, thus making the wastewater much clearer. It has therefore become common practice to call these sedimentation tanks "clarifiers". The first clarifier that the wastewater flows into is called a *primary* clarifier. We will discuss later the need for another clarifier after the biological treatment process. This second clarifier is called a secondary clarifier.

Clarifiers normally are either rectangular or circular. Primary clarifiers are usually designed to provide 1.5 to 2 hours detention time. Secondary clarifiers usually provide slightly more time.

Generally, the longer the detention time provided, the more removal of solids that takes place. In a tank with two hours detention time, approximately 60 percent of the suspended solids in the raw wastewater will either settle to the bottom or float to the surface and be removed. Removal of these solids will usually reduce the Biochemical Oxygen Demand (BOD) of the waste approximately 30 percent. The exact removal depends on the amount of BOD contained in the settled material.

All primary clarifiers, no matter what their shape, must have a means for collecting the settled solids (called sludge) and the floating solids (called scum). In rectangular tanks, sludge and scum collectors are usually wooden beams (flights) attached to endless chains. The collector flights travel on the surface, in the direction of the flow, conveying grease and floatable solids down to the scum trough to be skimmed off to the solids (sludge) handling facilities. The flights then drop below the surface and return to the influent and along the bottom, moving the settled raw sludge to the sludge hopper. The sludge is periodically pumped from the hopper to the sludge handling facilities.

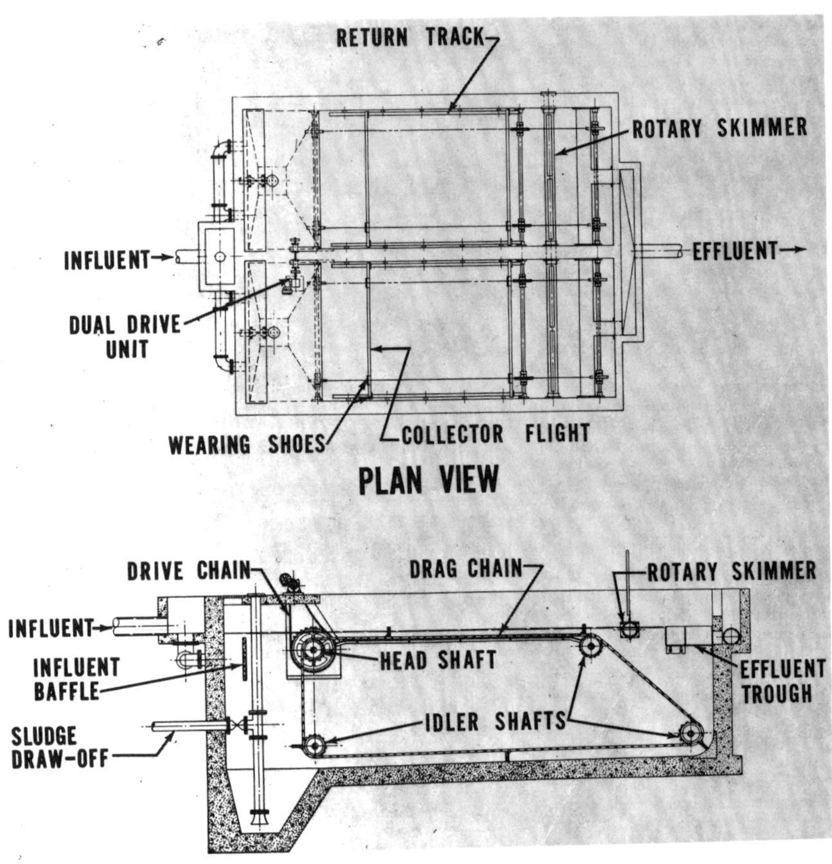

Figure 2. Typical Layout — Primary Tank (sectional view).

In circular tanks, scrapers or "plows", attached to a rotating arm, rotate slowly around the bottom of the tank. The plows push the settled sludge toward the center and into the sludge hopper. Scum is collected by rotating blade at the surface. As in the case of the rectangular tank, both scum and sludge are usually pumped to the solids or sludge handling facilities.

The clear surface water of the primary tank flows out of the tank by passing over a weir. The weir must be long enough to allow the treated water to leave at a low velocity; if it leaves at a high velocity, particles settling to the bottom or those already on the bottom may be picked up and carried out of the tank.

Trickling Filter

The trickling filter is one of the oldest and most dependable of the biological treatment processes. Most of these plants are removing 65 to 85% of the BOD and suspended solids present in the influent.

Table 2. Primary Treatment.

Clarifier Components	OPERATION			MAINTENANCE		
	Cleaning	Adjustments	Solids Removal	Lubrication	Replace Worn Parts	Painting
Collector Arms	Hose Down	1.	Daily	2.	3.	4.
Skimmers	Hose Down	1.	Daily	2.	3.	4.
Sludge Removal	—	To Match Digestion	Daily	—	—	—
Pumps/Valves Fittings	—	—	—	2.	3.	4.
Outlet Weirs	Hose Down	1.	Daily	—	3.	4.

Notes:

1. Periodically for alignment.
2. As Recommended by manufacturer.
3. At Recommendation of Manufacturer, Later by experience.
4. On a routine basis as a prentative measure.

Figure 3.

The trickling filter is a bed of 1–1/2 to 5-inch rock, slag, blocks or specially manufactured "media" over which settled wastewater from the primary clarifier is distributed. The settled wastewater is usually applied by an overhead rotating distributor and trickles over and around the media as it flows downward to the effluent collection channel. Since the media and the voids in between them are large (usually 2.5 to 4-inch diameter), and since the applied wastewater no longer has any large particles (they settled out in the clarifier), the trickling filter does not remove solids by a filtering action. It would be more correct to call the filter a biological contact bed or biological reactor since this is the function it performs. The filter bed offers a place for aerobic bacteria and other organisms to attach themselves and multiply as they feed on the passing wastewater. This process of feeding on, or decomposing waste is exactly the same as the process occurring in the stream when waste is discharged to it. In the trickling filter, however, the organisms use the oxygen which enters the waste from the surrounding air, rather than using up the stream's supply of dissolved oxygen. Thus the voids between the media must be large so sufficient oxygen can be supplied by circulating air.

The wastewater being distributed on the filter usually has passed through a primary clarifier, but it still contains approximately 70 percent of its original organic matter, which represents food for organisms. For this reason, a tremendous population of organisms develops on the media. This population continues to grow as more waste is applied. Eventually the layer of organisms on the media gets so thick that some of it breaks off (sloughs off) and is carried into the filter effluent channel. This material is normally called *humus*. Since it is principally organic matter, its presence in a stream would be undesirable. It is usually removed by settling in a *secondary clarifier*. Humus sludge from the secondary clarifier is usually returned to the primary clarifier to be resettled and pumped to the sludge handling facilities along with the "raw" sludge which settles out as previously described.

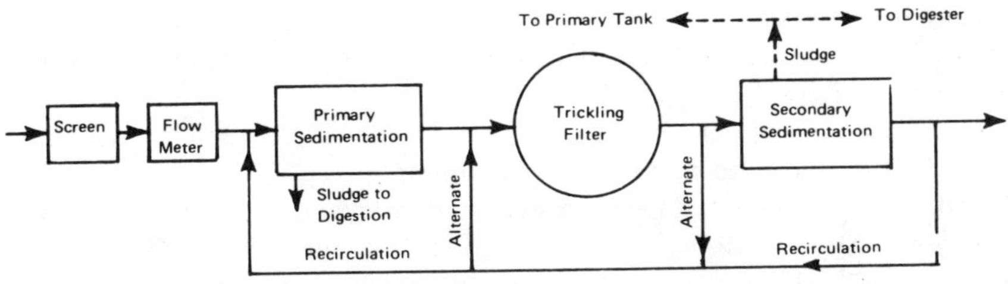

Figure 4. Trickling Filter Plant.

Table 3. Typical Trickling Filter Plant Lab Results.

	Test	Frequency	Location	Common Range
1.	Dissolved Oxygen	Daily	Prim. Effl.	1.0 - 2.0 mg/l
2.	Settleable Solids	Daily	Influent	5 - 15 ml/l
3.	pH	Daily	Influent Final Effl.	6.8 - 8.0 7.0 - 8.5
4.	Temperature	Daily	Influent	-----
5.	BOD	Weekly (Minimum)	Influent Prim. Effl. Final Effl.	150 - 400 mg/l 60 - 160 mg/l 15 - 40 mg/l
6.	Suspended Solids	Weekly (Minimum)	Influent Prim. Effl. Final Effl.	150 - 400 mg/l 60 - 150 mg/l 15 - 40 mg/l
7.	Chlorine Residual	Daily	Final Effl.	0.5 - 2.0 mg/l
8.	Coliform Bacteria	Weekly (Minimum)	Final Effl., Chlorinated	50 - 700/100 ml ----
9.	Clarity	Daily	Final Effl.	1 0 3 ft

Daily Operation

Once growth on the media has been established and the plant is in "normal operation," very little routine operational control is required. Careful daily observation is important. Items to be checked daily are:
1. Any indication of ponding
2. Filter files
3. Odors
4. Plugged orifices
5. Roughness or vibration of the distributor arms
6. Leakage past the mercury seal

Occasionally the underdrains should be checked for accumulations of debris in order to prevent stoppages.

Refer to the appropriate paragraphs in the following section on operational problems for procedures to correct these conditions.

Operation of clarifiers is interconnected with trickling filter operation. If the recirculation pattern permits, it is a good idea to return filter effluent to the primary clarifier. This is a very effective odor control measure. In some plants, increasing the recirculation rate will increase the hydraulic loading on the clarifier. *Be sure the hydraulic loading remains within the engineering design* limits. If the hydraulic loading is too low, septic conditions may develop in the clarifier, while excessively high loadings may wash solids out of the clarifier.

Recirculation during low inflow periods of the day and night may help to keep the slime growths wet, minimize fly development and wash off excessive slime growths. It may be necessary to reduce or stop recirculation during high flow periods to avoid clarifier problems from hydraulic overloading.

Secondary Treatment [2]

In many situations, primary treatment with the resulting removal of 40 to 60 per cent of the suspended solids and approximately 25 or 35 per cent of the BOD, together with the removal of floating material from the sewage, is adequate to meet requirements of the receiving water. However, if the accomplishment of primary treatment is not sufficient, there are two basic methods of secondary treatment available, trickling filters and activated sludge. There are several other methods which are used to a limited extent. These types of treatment employ biological growths to effect aerobic decomposition or oxidation of organic material into more stable compounds and provide for a higher degree of treatment than that accomplished by primary sedimentation alone.

When trickling filters and activated sludge both depend on aerobic biological organisms to effect decomposition, there is an operational difference. In filters, the organisms are attached to the filter medium and the organic material on which they do their work is brought to them. With activated sludge, however, the organisms are mirgrant and are carried to the organic matter in the sewage. In either case, successful operation involves the maintenance of aerobic environmental conditions favorable for the life cycle of the organisms, and control, over the amount of organic matter which they decompose. The organic matter is the food upon which these organisms live. If they are either over-fed or under-fed their efficiency is reduced.

Recirculation of final clarifier effluent dilutes influent wastewater and recirculated sludge improves slime development on the media. You should, by evaluating your own operating records, adjust the process to obtain the best possible results for the least cost. Use the lowest recirculation rates that will yield good results (but not causing ponding or other problems) to conserve power. Power costs are a large item in a plant budget. Also reduced hydraulic loadings mean better settling in the clarifiers, resulting in less chlorine usage in plants which disinfect the final effluent, since organic matter exerts a high chlorine demand. If filter effluent, rather than secondary clarifier effluent, is recirculated, the hydraulic loading on the secondary clarifier is not affected.

Table 4. Secondary Waste Treatment.

Type Process	Retention Time (includes clarifier) (Hrs)	Removal Efficiency (%)	
		Biochemical Oxygen Demand (BOD)	Suspended Solids (SS)
Trickling Filter	1 - 2	65 - 85	65 - 85
Activated Sludge	4 - 8	90 - 99	90 - 99
Extended Aeration	24	90 - 99	90 - 99
Oxidation Pond	Greater than 30 days	90 - 95	90 - 95

Activated Sludge

Another biological treatment unit that is used in secondary treatment, following the primary clarifier, is the aeration tank. When aeration tanks are used with the sedimentation process, the resulting plant is called an *activated sludge* plant. The activated sludge process is widely used by large cities and communities where land is expensive and where large volumes must be highly treated, economically, without creating a nuisance to neighbors. The activated sludge plant is probably the most popular biological treatment process being built today for larger installations or small package plants. These plants are capable of BOD and suspended solids reduction of up to 90 or 99%. The activated sludge process is a biological process, and it serves the same function as a trickling filter. Effluent from a primary clarifier is piped to a large aeration tank. Air is supplied to the tank by either introducing compressed air into the bottom of the tank and letting it bubble through the wastewater and up to the top, or by churning the surface mechanically to introduce atmospheric oxygen.

Aerobic bacteria and other organisms thrive as they travel through the aeration tank. With sufficient food and oxygen they multiply rapidly, as in a trickling filter. By the time the waste reaches the end of the tank (usually 4 to 8 hours), most of the organic matter in the waste has been used by the bacteria for producing new cells. The effluent from the tank, usually called "mixed liquor", consists of a suspension containing a large population of organisms and a liquid with very little BOD. The activated sludge forms a lacey network that captures pollutants.

The organisms are removed in the same manner as they were in the trickling filter plant. The mixed liquor is piped to a secondary clarifier, and the organisms settle to the bottom of the tank while the clear effluent flows over the top of the effluent weirs. This effluent is usually clearer than a trickling filter effluent because the suspended material in the mixed liquor settled to the bottom of the clarifier more readily than the material in a trickling filter effluent. The settled organisms are known as *activated sludge*.

Requirements for Control

Control of the activated sludge process is based on evaluation of and action upon several interrelated factors to favor effective treatment of the influent wastewater. These factors include:

1. Effluent quality requirements.
2. Wastewater flow, concentration, and characteristics of the wastewater received.
3. Amount of activated sludge (containing the working organisms) to be maintained in the process relative to inflow.
4. Amount of oxygen required to stablize wastewater oxygen demands and to maintain a satisfactory level of dissolved oxygen to meet organisms requirements.
5. Equal division of plant flow and waste load between duplicate treatment units (two or more clarifiers or aeration tanks).
6. Transfer of the pollution material (food) from the wastewater to the floc mass (solids or workers) and separation of the solids from the treated wastewater.
7. Effective control and disposal of inplant residues (solids, scums, and supernatants) to accomplish ultimate disposal in a nonpollutional manner.
8. Provisions for maintaining a suitable environment for the work force of living organisms treating the wastes. Keep them healthy and happy.

To prevent sludge bulking from occurring, the following items should be carefully controlled in an activated sludge plant:

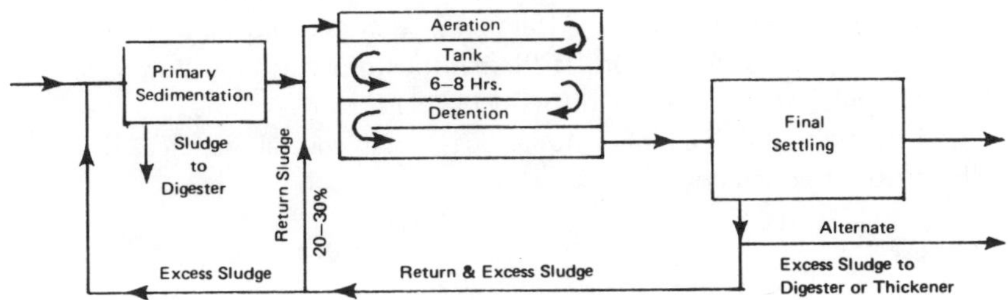

Figure 5.

Figure 6. Conventional Activated Sludge Plant.

*Table 5. Aeration Tank Capacities and Permissible Loadings**

PROCESS	PLANT DESIGN FLOW, MGD	AERATION RETENTION PERIOD, HOURS BASED ON DESIGN FLOW	PLANT DESIGN LOAD lb BOD/day	AERATOR LOADING lb BOD per day/lb MLSS	SLUDGE AGE, DAYS
Modified or "HIGH-RATE"	All	2.5 up	2000 up	1/1 (or less)	0.5 - 2.0
Conventional	To 0.5 0.5 to 1.5 1.5 up	7.5 6.0 to 7.5 6.0	To 1000 1000 to 3000 3000 up	1/2 to 1/4	3.5 - 7.0
Extended Aeration	All	24	All	As high as 1/10 to As low as 1/20	10 or Longer

* Recommended Standards for Sewage Works (10 State Standards), Great Lakes-Upper Mississippi River Board of State Sanitary Engineers, 1968 Edition, published by Health Education Service, P. O. Box 7283, Albany, New York 12224. Price, $1.00.

1. *Suitable Sludge Age.* Carefully review plant records and maintain a sludge age that produces the best quality effluent. Watch influent solids loadings, maintain desired level of solids in the aerator, and carefully regulate waste sludge rates. Generally, bulking may be cured by increasing the sludge age.

2. *Low DO.* Prevent low levels of DO from developing. Mixed liquor DO determinations are a quick, simple test; and if a DO probe is used, it gives a continuous reading. There is no valid excuse for low DO concentrations during normal conditions if sufficient oxygenation capacity is available, unless a slug of waste with an excessive oxygen demand is received.

3. *Short Aeration Period.* Bulking caused by the aeration period being too short is usually the result of a design problem, unless the operator has formed the habit of returning too high a volume of return sludge. This can be corrected by reducing the return sludge rate and thickening the return sludge solids concentration by coagulation, if necessary, thereby still returning the same number of organisms to meet the new food (waste) entering the aerator, but effectively reducing the total flow through the aerator and clarifier.

4. *Filamentous Growth.* Occurrence of filamentous growth may be caused by incorrect sludge age or nutritional differences, such as a shortage or abundance of nitrogen, phosphorus, or carbon. If filamentous growths are allowed to become well established, they create a difficult problem to overcome. Control may be achieved by maintaining a higher sludge age, and in special instances supplementing the nutrient deficiency.

Septic Sludge

Septic sludge may be produced when any type of sludge remains too long in such places as hoppers and channels. It is likely to cause a foul odor, rises slowly, and sometimes rises in clumps. Even small amounts can upset an aerator.

Septic sludge may occur in poorly designed or constructed hoppers, wet wells, channels, or pipe systems. This occurs when activated sludge is allowed to be deposited and anaerobic decomposition starts. Septic sludge deposits also may develop on the floor of the aerator due to insufficient air rates that are not keeping the tank completely mixed. A high solids load also can cause septic problems.

To effectively control septic sludge, aerators must be thoroughly mixed and sludge must be pumped frequently. In channels and pipelines, a velocity over 1.5 feet per second will prevent the formation of sludge deposits that could become septic.

Sludge going septic in the secondary clarifier may develop from four causes:
1. Return sludge rate too low, thus holding the solids in the final clarifier too long and allowing them to become septic.
2. Clarifier collection mechanism turned off, thus the sludge is not being moved to draw-off hopper.
3. Sludge draw-off lines plugged, obstructed, or used infrequently.
4. Return sludge pump off or a valve closed.

A good operator checks his system several times a day. In most new activated sludge plants the secondary clarifiers have air lift samplers or photocells to indicate sludge blanket level in the tank. Whenever the final clarifier sludge blanket level changes, an immediate investigation should be undertaken. In any of the cases above, the correction is quite obvious — restore suitable return sludge flow as soon as possible.

Toxic Substances

Toxicity causes inhibition or death of working organisms to produce system and effluent upsets. The operator has limited control over the cause. When this cause is identified, sludge wasting should be stopped immediately and all available solids returned to the aerator. Toxic materials such as heavy metals, acids, insecticides, and pesticides should never be dumped into a sewer system without proper control.

Rising Sludge

Rising sludge is not to be confused with bulking. The sludge settles and compacts satisfactorily on the bottom of the clarifier, but after settling it rises to the top of the secondary tank in patches or small particles the size of a pea. This is usually accompanied by a fine scum or froth (brown in color) on the surface of the aeration and secondary tanks.

Rising sludge is caused by denitrification or septicity and results from too long a detention time in the secondary clarifiers. The secondary clarifiers should be equipped with scum baffles and skimmers to prevent these solids from escaping in the plant effluent.

Denitrification is most common when the sludge age is high (extended aeration).

When this type of activated sludge flows from the aerator to the secondary clarifier or becomes short of oxygen, the organisms first use the available dissolved oxygen and the oxygen in the nitrates resulting in the release of nitrogen gas. Denitrification is an indication of good treatment, providing the sludge in the settleability test stays on the bottom of the cylinder for at least one hour, but floats to the surface in two hours. If it floats up too early in the settleability test, the sludge age should be reduced or the food-to-organism ratio should be increased. If the sludge stays down for an hour in the settleability test but problems are still present in the secondary clarifier, increase the return sludge rates to move the solids out of the clarifier at a faster rate.

Frothing

Aerator frothing has been a problem for some plants. There have been many theories presented on the cause, such as surfactants (detergents), polysaccharides, and over-aeration. Whatever the cause, there is a definite relationship between froth build-up on the aerator and the amount of suspended solids in the mixed liquor and air supply to the aerator.

For control:

1. Maintain higher mixed liquor suspended solids concentration.
2. Reduce air supply during periods of low flow while still maintaining DO.

Loading of Aeration Unit

As an operator, your most important concern is with the degree of treatment provided. This is normally expressed in % BOD removal. In order to maintain a consistent BOD removal, the BOD to solids loading (the pounds of BOD in one day's flow to the aerator when compared with the weight of solids under aeration (MLSS) of the aerator must remain within the limits for the type of treatment provided. These limits are shown in Table 6.

What are we talking about when we discuss loadings in terms of #BOD per 100# MLSS? The MLSS is the mixed liquor suspended solids or the concentration of solids in the aerator. To determine the total amount of solids, take the volume of your aerator $\frac{(L \times W \times D \times 7.5)}{1,000,000} \times MLSS$ in ppm $\times 8.34$

1. # Solids in aerator = Volume × ppm × 8.34 Divide your answer by 100 to get in terms of 100# MLSS.

   Find the total #BOD in one day's flow of sewage to the aeration unit.
2. #BOD — Average Daily Flow × ppm BOD × 8.34
3. Divide $(2) = \frac{\#BOD}{(1) \quad 100\# MLSS}$ = BOD loading to aerator

Example 1:   Aerator Volume = 250,000 gal.   MLSS − 2500 ppm
                    Raw BOD   = 200 ppm
                    ADF   = 1.0 mgd

(1a)   #Solids Under Aeration   = $\dfrac{250,000 \times 8.34 \times 2500}{1,000,000}$

                                         = 5210#
                                         = 52.10 (100# MLSS)

(2a)   #BOD = 200 × 8.34 × 1.0 = 1670#

(3a)   BOD to solids = $\dfrac{(2a)}{(1a)}$ = $\dfrac{1670\# \text{ BOD}}{52.10 \,(100\#\, \text{MLSS})}$ = 32# BOD/100# MLSS

Note: This is in Activated Sludge Range.

We have just shown the calculations of what would be considered a normal strength sewage.

Now let us look at a weak sewage, using the same aerator, MLSS concentration, and flow. This would be similar to a plant which is receiving a weak sewage which is also then settled in a primary settling unit to get a strength of about 100 ppm BOD.

Example 2:

(1b)   # Solids in Aerator — Same as (1a) — 5210#

(2b)   #BOD = 100 ppm × 8.34 × 1.0 mgd = 835#

(3b)   BOD to solids = $\dfrac{(2b)}{(1b)}$ = $\dfrac{835\# \text{ BOD}}{52.10 \,(100\#\, \text{MLSS})}$ = 16# BOD/100# MLSS

Note: This loading is near the Extended Air Range.

Now let us look at a strong sewage, using the same aerator, solids and flow. The strength of waste will be 400 ppm.

Example 3:

(1c)   # Solids in Aerator = Same as (1a)

(2c)   = 400 ppm × 8.34 × 1.0 = 3340# BOD

(3c)   BOD to Solids = $\dfrac{(2c)}{(1c)}$ = $\dfrac{3340\# \text{ BOD}}{52.10 \,(100\#\, \text{MLSS})}$ = 64# BOD/100# MLSS

Note: This is beyond the range of Activated Sludge and toward Modified High Rate Activates Sludge. Therefore, you would expect to get a reduced degree of treatment.

Now let us look at each example from an operation point of view. In *Example 1*, we note that if our MLSS are increased, this decreases our loading and approaches the Extended Air Process.

Loading = $\dfrac{\text{BOD (Constant)}}{\text{MLSS Solids}}$ — As solids increase, loading will decrease

With this, the air requirements needed to maintain the processes will increase.

Using this same example, if the solids are decreased, the loading increases, causing the process to approach Modified High Rate Activated Sludge, resulting in a decrease in the air requirements and also a decrease in BOD removal.

With this in mind, let us look at *Example 2*. This is an underloaded plant. It would be best to increase the solids and thus decrease the loading to approximately 10# BOD/100# MLSS. This increases the air requirement, but if your plant were designed on normal design sewage strength (200 ppm BOD), sufficient air would be available. Also, all sludge could be kept in the system, with only periodic wasting of sludge, because the plant is operating as Extended Aeration.

Example 3. The solids would have to be increased to bring the loading to below 50# BOD/100# MLSS to insure 90% + BOD removal. The air requirements would be greater, and this could be the limiting factor of operation.

Table 6.

Process	BOD to Solids Loading (#BOD/100# MLSS)	Removal (%)	Air Requirements ** Mechanical (# O_2/# BOD_5) Applied	Diffused (cf/#BOD)
Extended Aeration	5 10 (Normal Design) 15	90+	2.4	2000 1750 1300
Activated Sludge and Contact Stabilization*	25 35 (Normal Design) 50	90+	1.0	1000 750
Modified (High Rate) Activated Sludge	100 250 500 1000	75 to 60		520 350 300 275

* Solids loadings and air requirements, when pertaining to contact stabilixation, must include the total solids under aeration in both the mixing (contact) and the reaeration (stabilization) tanks.

** Air requirements indicated are for process needs only, and do not include excess air for operation of air lift pumps, or for turbulence, etc.

Air Requirements: [1]

Although many factors may effect the air requirements of activated sludge plants, by far the most important is the ratio between the weight of solids carried under aeration and the weight of BOD input to the aeration tanks. There appears to be a direct relationship between this ratio and the cubic feet of air consumed per pound of BOD removed, which is illustrated by Table 6. Since that ratio also has an important bearing on the percentage reduction in BOD achieved by the process,

[1] "Suggested Design Criteria for the Activated Sludge Process", T. R. Hazeltine.

Typical results of lab tests for an activated sludge plant are provided to assist in the evaluation of lab results and plant performance. Remember that every plant is different and is influenced by different conditions.

Table 7. Typical Lab Results for an Activated Sludge Plant.

Test	Location	Common Range	
COD	Influent	300 – 700	mg/l
	Primary Effl.	200 – 400	mg/l
	Final Effl.	30 – 70	mg/l
	(Conv. Act. Sl.)		
BOD	Influent	150 – 400	mg/l
	Primary Effl.	100 – 280	mg/l
	Final Effl.	10 – 20	mg/l
	(Conv. Act. Sl.)		
SUSPENDED	Influent	150 – 400	mg/l
SOLIDS	Primary Effl.	60 – 160	mg/l
	Mixed Liquor	1000 – 4500	mg/l
	Return Sludge	2000 – 10,000	mg/l
	Final Effl.	10 – 20	mg/l
	(Conv. Act. Sl.)		
DISSOLVED	Mixed Liquor	2 – 4	mg/l
OXYGEN	Final Effl. (Outfall)	2 – 6	mg/l
CHLORINE RESIDUAL (30 min.)	Final Effl.	0.5 – 2.0	mg/l *
COLIFORM GROUP BACTERIA, MPN	Final Effl. (Chlorinated)	23 – 700/100 ml	
CLARITY (Secchi Disc)	Final Effl.	3 – 8	ft
pH	Influent	6.8 – 8.0	
	Effluent	7.0 – 8.5	

* Regulatory agencies normally specify a chlorine residual remaining after a certain time period.

Although many factors may effect the air requirements of activated sludge plants, by far the most important is the ratio between the weight of solids carried under aeration and the weight of BOD input to the aeration tanks. There appears to be a direct relationship between this ratio and the cubic feet of air consumed per pound of BOD removed, which is illustrated by Table 6. Since that ratio also has an important bearing on the percentage reduction in BOD achieved by the process, there can be no single normal air requirement per pound of BOD input to the aeration tanks.

Waste Treatment Ponds

Shallow ponds (three to five feet deep) are often used to treat wastewater and other wastes instead of, or in addition to conventional waste treatment processes. Wastes which are discharged into ponds are treated or stabilized by several natural processes acting at the same time. Heavy solids settle to the bottom where they are decomposed by bacteria. Lighter suspended material is broken down by bacteria in suspension.

Dissolved nutrient materials, such as nitrogen and phosphorous, are utilized by green algae which are actually microscopic plants floating and living in the water. The algae utilize carbon dioxide (CO_2) and bicarbonates to build body protoplasm. In so growing they need nitrogen and phosphorous in their metabolism much as land plants do. Like land plants, they release oxygen and some carbon dioxide as waste products.

In recent years, ponds have become more popular as treatment facilities. Extensive studies of their performance have led to a better understanding of the natural processes by which ponds treat wastes. Information is also available which can help operators to regulate the pond processes for efficient waste treatment.

Ponding of wastewater as a complete process offers the following advantages for smaller installations, provided land is not costly and the location is isolated from residential, commercial and recreational areas:

1. Does not require expensive equipment.
2. Does not require highly trained operating personnel.
3. Is economical to construct.
4. Provides treatment that is equal to or superior to some conventional processes.
5. Makes a satisfactory short-term method of treating wastewater on a temporary basis until a permanent plant can be constructed.
6. Is adaptable to fluctuating loads.
7. Is probably the most trouble-free of any treatment process when utilized correctly, provided a consistently high quality effluent is not required.

Because ponds are deceptively simple, they are probably neglected more than any other type of wastewater treatment process. Many of the complaints that arise from ponds are the result of neglect or poor housekeeping. Following are listed the day-to-day operational and maintenance duties that will help to insure peak treatment efficiency and to present your plant to its neighbors as a well-run waste treatment facility.

Scum Control

Scum accumulation is a common characteristic of ponds and is usually the greatest in the spring of the year when the water warms and vigorous biological activity resumes. Ordinarily, wind action will dissipate scum accumulations and cause them to settle; however, in the absence of wind or in sheltered areas, other means must be used. If scum is not broken up, it will dry on top and become crusted. It is not only more difficult to break up then, but a species of blue-green algae is apt to become established on the scum which can give rise to disagreeable odors. If scum is allowed to accumulate, it can reach proportions where it cuts off a significant amount of sunlight from the pond.

Rafts of scum cause a very unsightly appearance in ponds and can quite likely become a source of botulism that will have a devastating effect on waterfowl and shore birds which may be attracted to the facility.

Many methods of breaking up scum have been used, including agitation with garden rakes from the shore, jets of water from pumps or tank trucks, and the use of outboard motors on boats in large ponds. Scum is broken up most easily if it is attended to promptly.

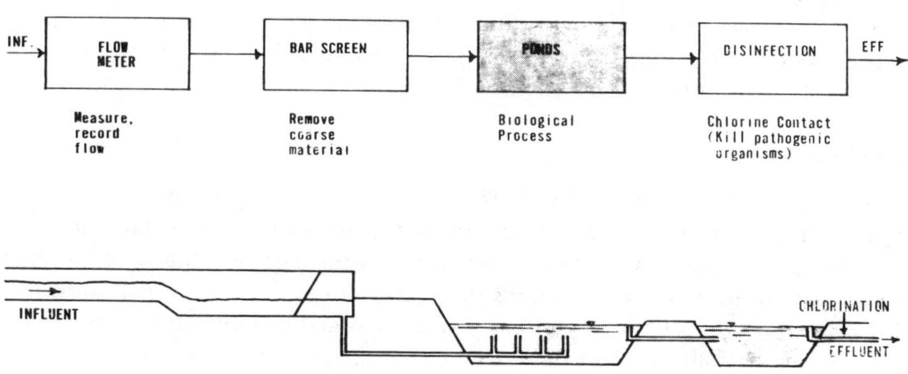

Figure 7. Typical Plant; Ponds Only.

Odor Control

It is probably inevitable that, at some time, odors will come from a wastewater treatment plant no matter what kind of process is used. Most odors are caused by overloading or poor housekeeping practices and can be remedied by taking corrective measures. However, there are times, such as when unexpected shutdowns occur, that plant processes may be upset and cause odors. For these unexpected occurrences it is strongly advised that a careful plan for emergency odor control be available. Odors usually occur during the spring warmup in colder climates because biological activity is reduced during cold weather.

For ponds, recirculation from aerobic units, the use of floating aerators, and heavy chlorination should be considered as means to reduce odors. Recirculation from an aerobic pond to the inlet of an anerobic pond (1 part recycle flow to 6 parts influent flow) will reduce or eliminate odors. Usually floating aeration and chlorination equipment are too expensive to have setting idle waiting for an odor problem to develop. Odor masking chemicals also have been promoted for this purpose and have some uses for concentrated specific odor sources. However, in almost all cases, process procedures of the type mentioned previously are preferable. In any event, waiting until the emergency arises before planning for odor control is poor procedure. Often several days are needed to receive delivery of materials or chemicals if they are required. Try to have possible alternate methods of control ready to go if they are needed.

In some areas, sodium nitrate has been added to ponds as a source of oxygen to prevent odors. To be effective, sodium nitrate must be dispersed throughout the water in the pond. Once mixed in the pond it acts very quickly because many common organisms (facultative groups) may use the oxygen in nitrates instead of dissolved oxygen. Liquid sodium hydrochloride or chlorine solution is a faster acting solution, but not necessarily the best chemical because it will interfere with biological stabilization of the wastes.

Weed Control

Weed control is an essential part of good housekeeping and is not a formidable task with modern herbicides and soil sterilants. Weeds around the edge are most objectionable because they allow a sheltered area for mosquito breeding and scum accumulation. In most average ponds there has been little need for mosquito control when edges are kept free of weed growth. Aquatic weeds, such as tules, will grow in depth shallower than three feet, so an operating pond level of at least this depth is necessary. Tules may emerge singly or well scattered but should be removed promptly by hand as they will quickly multiply from the root system. Weeds also can hinder pond circulation.

Insect Control

Mosquitoes will breed in sheltered areas of standing water where there is vegetation or scum to which the egg rafts of the female mosquito can become attached. These egg rafts are fragile and will not withstand the action of disturbed water surfaces such as caused by wind action or normal currents. Keeping the water edge clear of vegetation and keeping any scum broken up will normally give adequate control. Shallow, isolated pools left by a receding pond level should be drained or sprayed with a larvacide.

Any of several minute shrimp-like animals may infest the pond from time to time during the warmer months of the year (March — November). These predators live on algae and at times will appear in such numbers as to almost clear the pond of algae. During the more severe infestations there will be a sharp drop in the dissolved oxygen of the pond, accompanied by a lowered pH. This is a temporary condition because the predators will outrun the algae supply, and there will be a mass die-off of insects which will be followed by a rapid greening up of the pond again.

Ordinarily there should be no great concern about these infestations because they soon balance themselves; however, in the case of a heavily loaded pond, a sustained low dissolved oxygen content may give rise to noxious odors. In that event any of several commercial sprays can be used with excellent control. Dibrom-8 has been used with good results.

Chironomid midges are often produced in wastewater ponds in sufficient numbers to be serious nuisances to nearby residential areas, farm workers, recreation sites, and industrial plants. When emerging in large numbers they may also create traffic hazards. At present the only satisfactory control is through the use of insecticides such as parathion, Abate, Sursban, and Fenthion. Control measures are time consuming and may be difficult, particularly if there is a discharge to a receiving stream. If possible, lower the level in the pond enough to contain a day's inflow before applying an insecticide. Holding the insecticide for at least one day will kill more insects and reduce the effect of the insecticide on receiving waters. For better results, insecticides should be applied on a calm day and any recirculation pumps should be stopped.

Levee Maintenance

Levee slope erosion caused by wave action is probably the most serious maintenance problem. If allowed to continue, it will result in a narrowing of the levee crown which will make accessibility with maintenance equipment most difficult.

If the levee slope is composed of easily erodable material, the only long-range solution is the use of bank protection such as stone riprap or broken concrete rubble.

Levee tops should be crowned so that rain water will drain over the side in a sheet flow rather than flowing a considerable distance along the levee crown and gathering enough flow to cause erosion when it finally spills over the side and down the slope.

If the levees are to be used as roadways during wet weather, they should be paved or well graveled.

Headworks and Screening

It is important to clean the bar screen as frequently as possible. The screen should be visited at least once or twice a day with more frequent visits during storm periods. Screenings should be disposed of daily in a sanitary manner, such as by burial, to avoid odors and fly breeding.

Many pond installations have grit chambers at the headworks to protect raw wastewater lift pumps or prevent plugging of the influent lines. There are many types of grit removal equipment. Grit removed by the various types of mechanical equipment or by manual means will usually contain small amounts of organic matter and should therefore be disposed of in a sanitary manner. Disposal by burial is the most common method.

SLUDGE HANDLING AND DISPOSAL

Introduction [3]

Suspended solids are usually present in the influent to a municipal wastewater treatment plant at levels of 100 to 300 mg/l. Additional suspended solids are generated during the various waste water treatment processes. Some result from biological processes and others are generated by chemical precipitation.

Sludge is a broad term used to describe the various aqueous suspensions of solids encountered during treatment. The nature and concentration of the solids control the processing characteristics of the sludge. Grit, screenings, and scum are not normally considered as sludge, therefore are not discussed in this chapter.

The following processes are used in sludge handling and disposal:

a. Thickening
b. Stabilization
c. Conditioning
d. Dewatering
e. Heat drying
f. Final Disposal

In classifying and describing sludge processing methods, the potential of a process to accomplish more than one task must be taken into account. Accordingly, the nomenclature attempts to recognize that four of the major categories (Thickening, Conditioning, Dewatering, and Reduction) have primary as well as secondary objectives.

Sludge Thickening (Blending)

The term thickening, herein, will be used to describe an increase in solids concentration, whether it occurs as the objective of a separate process, or as a secondary effect of a process provided essentially for a different purpose. Thickening Methods (Blending) are as follows:

a. Gravity
b. Flotation
c. Centrifugation

Recognition of the need to uniformly blend or combine the two principal types of wastewater sludges (primary and excess activated), and to keep them combined in plants where joint processing is practiced, is not as widespread as it should be. Normally, sludge blending can best be accomplished in a separate sludge thickening process.

Sludge Stabilization (Reduction)

Sludge stabilization processes are aimed at converting raw (untreated) sludges into a less offensive form with regard to odor, putrescibility rate, and pathogenic organism content. Major types of processes are:

a.   Anerobic digestion
b.   Aerobic digestion
c.   Lime treatment
d.   Chlorine oxidation
e.   Heat treatment
f.   Composting

Some discussion of each term follows:

Anerobic and aerobic digestion involve the biological stabilization of sludge through partial conversion of putrescible matter into liquid, dissolved solids, and gaseous by-products with some destruction of pathogens. These processes also reduce the amount of dry sludge solids. Consequently, these processes result in stabilization and in solids reduction or conversion.

Lime treatment and chlorine oxidation both control odor and reduce pathogens without significantly reducing sludge solids.

Heat treatment kills pathogenic organisms. In addition, putrescible organic matter is substantially dissolved and appears in the cooking liquor from subsequent decantation or dewatering.

Composting is an aerobic process involving the biological stabilization of sludge. It provides organic solids, pathogen, and odor reduction.

Sludge Conditioning is pretreatment of a sludge to facilitate removal of water in a thickening or dewatering process. Methods are as follows:

a.   Chemical (Inorganic and Organic)
b.   Elutriation
c.   Heat treatment

Chemical methods involve the use of inorganic or organic flocculants to promote formation of a porous, free-draining cake structure. In this way, the flocculants improve sludge dewaterability, alter sludge blanket properties, and improve solids capture. In dewatering, flocculants increase the degree of solids capture both by destabilization and agglomeration of fine particles and facilitate cake formation. The resultant cake becomes the true filter media. In thickening processes, the flocculants promote more rapid phase separation, higher solids contents, and a greater degree of capture.

Elutriation is the process of washing the alkalinity out of anaerobically digested sludge to decrease the demand for acidic chemical conditions and to improve settling and dewatering characteristics. When used with primary sludge, the process is cost-effective and does not create undesirabley effects. When elutriation is used in a plant which combines primary and excess activated sludge prior to digestion, the mixed sludge fractionates during the elutriation process, producing a highly polluted elutriate. The process has been criticized because this elutriate was bypassed into the plant effluent at some plants. However, use of flocculants in elutriation can eliminate the problem of the polluted elutriate.

Heat treatment, herein, refers to the pressure cooking of sludges in such a manner that little sludge oxidation occurs. The Porteous, Farrer, Zurn, and some

Zimpro systems fall into this category. Thus, heat treatment is distinct from wet air oxidation which generally involves higher temperatures and pressures, with air injection to promote a major degree of sludge oxidation.

Table 8. Conditioning Methods and Purposes.

Conditioning Method	Unit Process	Function
Polymer Addition	Thickening	Improve loading rate, degree of concentration, and solids capture
Polymer Addition	Dewatering	Improve production rate, cake solids content, and solids capture
Inorganic Chemical Addition	Dewatering	Improve production rate, cake solids content, and solids capture
Elutriation	Dewatering	Decrease acidic chemical conditioner demand and increase degree of concentration
Heat Treatment	Dewatering	Eliminate or decrease chemical use, improve production rate, cake solids content, and stabilization. Some conversion may also occur
Ash Addition	Dewatering	Provide improved cake release from belt type vacuum filters and facilitate filter pressing. It can also result in higher filter yields and reduced chemical requirements

Dewatering Methods

Any process which removes sufficient water from sludge so that its physical form is changed from essentially that of a fluid to that of a damp solid, is a dewatering process. Methods used in dewatering are best described by the equipment employed and some major types are:

a.   Rotary vacuum filters
b.   Centrifuges
c.   Drying beds
d.   Filter presses
e.   Horizontal belt filters
f.   Rotating cylindrical devices
g.   Lagoons

Heat Drying of Sludge

Sludge drying processes involve the application of heat to evaporate sufficient moisture and render the sludge dry to the touch and relatively free flowing. It is normal practice to conserve energy by dewatering the sludge prior to heat drying. Principal types of dryers are:

a.   Multiple hearth
b.   Flash dryers

c.   Tray dryers
d.   Spray dryers

Sludge Reduction

Sludge reduction, as defined here, pertains to processes which primarily yield a major reduction in the volatile sludge solids. Principal methods of sludge reduction are:

a.   Incineration
b.   Wet air oxidation
c.   Pyrolysis

Final Disposal Methods

Final or ultimate disposal refers to the disposition of sludge in liquid, cake, dried, or ash form, as a residue to the environment. Principal methods are:

a.   Cropland application
b.   Land reclamation
c.   Power generation (with solid waste)
d.   Sanitary landfill
e.   Ocean disposal

The first three methods are also utilization procedures. In instances where sanitary landfills are used for purposes of topographic modification, this also could be construed as utilization.

Sequence and Functions of the Unit Processes

The following figure summarizes the purposes and sequence of unit processes of wastewater sludge treatment.

FINAL DISPOSAL PROCESSES

Methods, Functions, and Occurrences

The need for proper treatment of all residual sludge streams is well recognized and should be provided for in plant design. Lakes and streams are no longer acceptable for sludge disposal. The methods of final disposal can be broadly categorized as disposal or utilization procedures. Disposal procedures include landfills and ocean dumping as described.

The landfill is one of the major methods for final disposal of sludge and incinerator ash. The amount of sludge that is dumped in the ocean is decreasing because of regulations.

Utilization procedures are receiving increasing attention and are described in Table 10. Cropland application and land reclamation are the major sludge utilization methods. The EPA municipal inventory shows that landfill and land spreading of

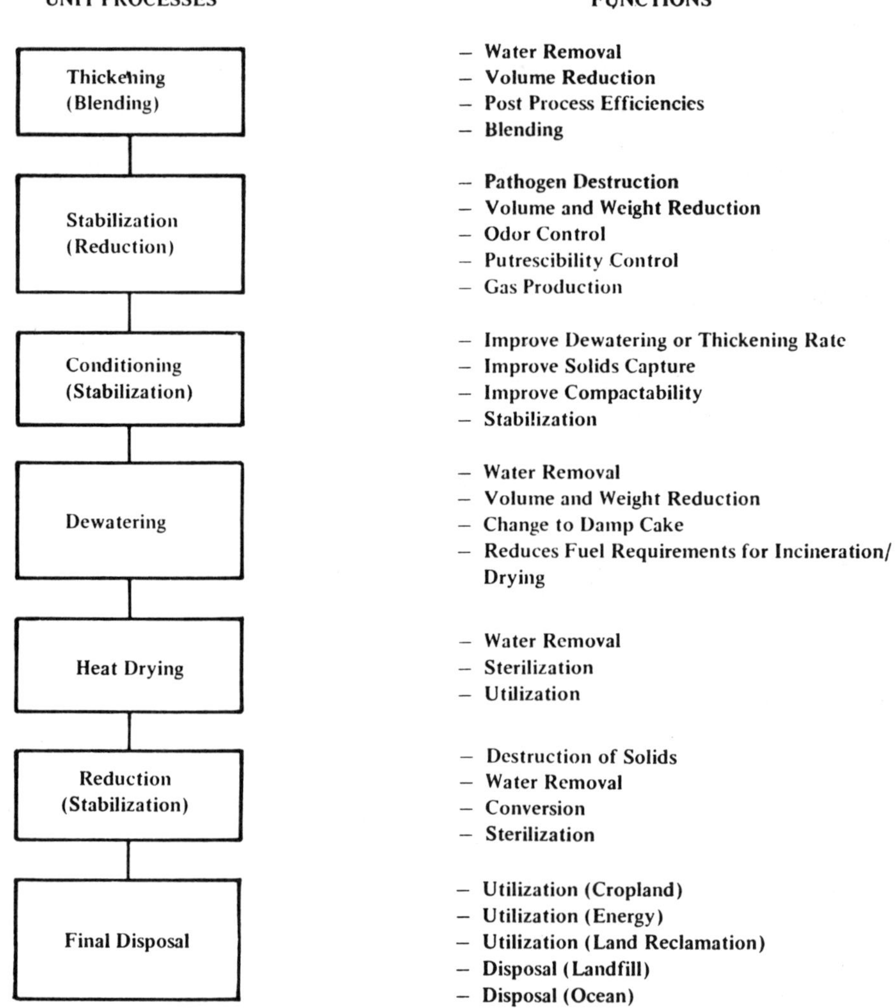

Figure 8. Enumeration of Sludge Treatment Processes and Their Functions.

sludge are used in 50 per cent of the U.S. installations. Proceedings of recent symposia on this topic are available and contain much useful information. Other methods of synthesizing or retrieving useful products by treating sludge with chemicals and/or heat are being studied.

Table 9. Final Disposal Methods.

Disposal Procedure	Principal Sludge Form	Main Constraints
Sanitary Landfill	Dewatered cake or ash (Stabilized)	Gas leachate, and runoff control; land availability
Ocean Dumping	Liquid (Thickened)	Oceanic and shoreline pollution

Table 10. Sludge Utilization Methods.

Utilization Procedure	Sludge Form	Main Constraints
Cropland Application	Liquid, cake dried or compost	Application rate, unsatisfactory sludge
Land Reclamation	Liquid or Dewatered	Application rate, unsatisfactory sludge, availability of land

Selection of Method of Final Disposal

In developing an optimum conceptual design for a wastewater treatment plant, it is increasingly apparent that determination of the method of final sludge disposal is a major consideration. The method of final disposal determines the acceptable form of the sludge residue and thus influences the choice of both sludge and liquid unit processes to be employed.

In selecting a final disposal or utilization process, the method chosen should be in accordance with local, state, interstate, and federal requirements. While no sludge residues, grit, ash, or other solids should be discharged into the plant effluent or receiving waters, the procedure chosen should also not result in any significant degradation of surface or groundwater; air or land surfaces.

Since present indications are that ocean disposal could be banned, designs incorporating sludge disposal to the ocean are questionable. It is desirable that methods chosen do not cause a health hazard or a nuisance condition, therefore it is essential that sludge be stabilized prior to spreading on land. The acceptable form of sludge for disposal or utilization is partially determined by the sludge treatment method. The sludge may be in the form of liquid, dewatered cake, incinerator ash, compost product, or dried powder.

Sanitary Landfill

Stabilized sludge containing no free water can be satisfactorily disposed in a sanitary landfill either alone or in a mixture with municipal solid waste. A sanitary landfill must be managed so that wastes are systematically deposited and covered with earth to control environmental impacts within defined limits. This distinguishes a sanitary landfill from an uncontrolled dumping operation. Sludges and solid wastes have in the past been disposed of in dumps not meeting proper landfill specifications. The placement of incinerator ash or stabilized sludge cake in a sanitary landfill can be an acceptable procedure when adequate land is available and site location and operational precautions prevent the creation of nuisance conditions or health hazards. Prior to placing sludge in a landfill it should be sufficiently dewatered to minimize the quality of free water present. Leachate and runoff from a sanitary landfill should be minimized and when necessary collected and suitably treated to prevent pollution of ground and surface waters. Therefore, sound engineering judgment dictates that sanitary landfills not be located in an existing flood plain.

Adequate monitoring of any land application or landfill site is essential. This plan must be specifically designed for applicable local conditions and should include monitoring groundwater observation wells, surface water, sludge and soils for heavy metals, persistent organic, pathogens, and nitrates. Human food chain products grown in sludge aided soils should also be monitored for heavy metals, persistent organics, and pathogens.

Three methods have been proposed to be used (either concurrently or separately) to minimize aerosols:
1.   low pressure spray
2.   downward directed spray, and
3.   large droplet spray
The potential for erosion should be checked before using these methods. Application methods other than spraying minimize aerosols. At many reclamation sites, however, spraying is the only practical method. When spraying is used, spraying methods should be selected which will minimize aerosol formation and wise buffer zones should be used. Mosquito vectors can be controlled by eliminating ponding on the site. This starts with properly grading the site and is made effective by maintenance to correct minor ponding locations and prevent others from forming.

Fruits and vegetables grown on wastewater irrigated or sludge-amended soils can be surface-contaminated with pathogenic microorganisms. The viability of these pathogens is extremely variable and may be from a few hours to several months. Also, among the factors which influence the survival of pathogens in the soil and on vegetation are:
a. type of organism
b. temperature-lower temperature increases viability
c. moisture-longevity is greater in moist soils than in dry soils
d. type of soil — neutral, high moisture holding soils favor survival
e. organic matter — the type and amount of organic matter present may serve as a
   food or energy source to sustain the microorganisms
The fact that some pathogens can survive the sludge digestion process, and remain viable in the soil for periods up to several months has prompted the general

recommendation that liquid sludges should not be applied to root crops or crops intended for human consumption in the raw form.

Pastureland and farmland used to grow forage crops are frequently used as land disposal sites, a practice which appears to present little problem from the standpoint of disease transmission via livestock grazing on such fields.

Use of Sludge on Agriculture Land

The application of sludge to farmland is a very popular utilization method because it can be both economical and simple. Limitations include heavy metals occasional public resistance and the unavailability of suitable land. The popularity of sludge spreading for crop production and soil improvement has increased significantly over the last ten years.

Metal content of sludge varies widely. The metals of most concern are zinc, copper, nickle, cadmium and lead. Zinc and copper are micronutrients that may enhance crop quality when sludge is spread thinly. The heavy metal content of sludge can vary widely from treatment plant to treatment plant.

Pathogen control is of importance because of the possible exposure directly to sludge in the handling and application steps and in the food chain. Although anaerobic digestion reduces the pathogen content of the sludge, a significant number of the pathogens may survive the process.

Aerosols which could contain pathogenic microorganisms may be present in the air over a landspreading site. Where spray irrigation is used to distribute the sludge, the potential for aerosols is increased.

Advanced Treatment Processes

Due to the requirements of the 1972 amendments to the Federal Water Pollution Control Laws and other regulations, sewage has been required to be treated to a greater degree than can be achieved by the primary and secondary treatments as described heretofore. In primary and secondary treatment methods some measurable BOD and some suspended solids as well as phosphorus, nitrogen and other organic constituents are left in the treatment plant effluent. Now more, if not most, of the ingredients must be removed by some additional treatment method. The amount removed depends upon the required or established quality of the waters receiving the treated effluent. It appears that the effluent quality standards will be continually upgraded.

To cope with this private and governmental research is continuously searching for new wasste treatment processes.

The terms, tertiary treatment and advanced waste treatment, have been applied to waste treatment beyond secondary that produces a better quality effluent than primary and secondary. On the other hand the term, "advanced waste treatment" can be applied to an entirely new concept of waste treatment plant, using a variety of techniques to attain the high degree of waste treatment required.

For purposes of this section, the term "tertiary treatment" will be applied to treatments that "polishes" the effluent from primary and secondary plants. Such methods include among others removal of solids by filtration, removal of nutrients by filtration and additional chemical sedimentation.

In solids removal by filtration, several types of filter media and types of filters can be used. Dual media and multi media filter beds have been developed which remove solids not removed by secondary treatment.

Diatomaceous earth filtration and ultra screening also can be applied to the treatment of secondary effluent. There are several patented processes of both of these methods readily available on the market.

In the removal of nutrients by filtration granular activated carbon beds can be used to adsorb organic compounds dissolved in the secondary effluent. These organic compounds can include phosphorous and nitrogen. Several companies are producing and marketing various activated carbons for such use. Carbon regeneration burns off the organics. The carbon can be regenerated with very little carbon loss.

Ion exchange processes are sometimes used in combination with sand filtration.

Chemicals can be used on secondary effluent and remove suspended solids through sedimentation. By properly controlled application of chemical coagulation phosphates can be removed. Coagulates used include lime, ferric chloride, ferric sulfate, aluminum sulfate, and ferrous sulfate.

Several waste treatment manufacturers have developed complete treatment systems to treat waste to a very high degree of treatment. These systems include variations of the treatment processes discussed herein and others completely without any biological aspect, including only physical and chemical treatment.

The following table is a brief evaluation of wastewater treatment processes giving the application, advantages and disadvantages. Several tertiary treatment processes not mentioned in this chapter are listed in this table.

BIBLIOGRAPHY

1. *Operation of Wastewater Treatment Plants* prepared for EPA under Grant No. 5TT1-WP-16-03, by Sacramento State College in cooperation with the California WPCA (1970).
2. *Manual of Instruction for Sewage Treatment Plant Operators*, New York State Department of Health.
3. *Process Design Manual for Sludge Treatment and Disposal* USEPA Technology Transfer, October 1974.
4. Paper on *Loading of Aeration Unit* by James E. Santarone, Division of Waste Water, bureau of Sanitary Engineering, Florida State Board of Health. Presented at Region 3 Short School, Brevard County (1966).

Table 11. Evaluation of Wastewater Treatment Processes.

Process	Application	Advantages	Disadvantages
1. PRETREATMENT			
a. Equalization	Waste streams with high variability	1. Dampens waste variations. 2. Reduces chemical requirements. 3. Dampens peak flows, reduces treatment plant size.	1. Need large land areas. 2. Possible septicity, requiring mixing and/or aeration equipment.
b. Neutralization	Waste streams with extreme pH values	1. Provide the proper conditions for biological, physical chemical treatment. 2. Reduces corrosion and scaling.	1. May generate solids. 2. Sophisticated equipment and instrumentation. High initial equipment costs.
c. Temperature Adjustment	Waste streams with extreme temperatures	Provides the proper conditions for biological treatment.	
d. Nutrient Additions	Nutrient deficient wastes	Optimizes biological treatment.	
2. PRIMARY TREATMENT			
a. Screening	Waste streams containing large solids (wood, rags, etc.)	1. Prevents pump and pipe clogging.	1. Maintenance required to prevent screen plugging, ineffective for sticky solids.
b. Grit Removal	Waste streams containing significant amounts of large, heavy inorganic solids.	1. Lowers maintenance costs, erosion 2. Reduces the solids loading to other treatment units.	Solids to be disposed of are sometime offensive.
c. Sedimentation	Waste streams containing settleable suspended solids	1. Reduces inorganic and organic solids loadings	1. Possible septicity and odors.

(continued)

Table 11. (continued)

		to subsequent biological units. 2. By far the least expensive and most common method of solid-liquid separation. 3. Suitable for treatment of a wide variety of wastes 4. Requires simpler equipment and operation. 5. Demonstrated reliability as a treatment process.	2. Adversely affected by variations in the nature of the waste. 3. Moderately large area requirement.
d. Dissolved-Air Flotation	Waste streams containing oils, fats, suspended solids and other flotable matter. Can be used for either clarification or thickening.	1. Removes oils, greases, and suspended solids. 2. Less tank area than for a sedimentation tank. 3. Higher concentration of solids than for sedimentation. 4. Satisfies immediate oxygen demand. Maintains aerobic conditions.	1. High initial equipment costs. 2. Sophisticated equipment and instrumentation. 3. High power and maintenance cost.
3. SECONDARY TREATMENT a. Activated Sludge (aeration and secondary sedimentation)	Biologically treatable organic wastes.	1. Flexible — can adapt to minor pH, organic and temperature changes. 2. Produces high quality effluent-90% BOD and suspended solids removal. 3. Small area required. 4. The degree of nitrification is controllable. 5. Relatively minor odor problems.	1. High operating costs (skilled labor, electricity, etc). 2. Generates solids requiring sludge disposal 3. Some process alternatives are sensitive to shock loads, and metallic or other posions 4. Requires continuous air supply.

Process	Application	Advantages	Disadvantages
b. Aerated Lagoon	Biologically treatable organic wastes.	1. Flexible — can adapt to minor pH, organic, and temperature waste changes. 2. Inexpensive construction. 3. Minimum attention. 4. Moderate effluent (50-75% BOD Removal).	1. Dispersed solids in effluent. 2. Affected by seasonal temperature variations. 3. Operating problems (ice, solids settlement, etc). 4. Moderate power costs. 5. Large area required. 6. No color reduction.
c. Oxidation Pond	Biologically treatable organic	1. Low Construction costs. 2. Non-skilled operation. 3. Moderate quality effluent (70-85% BOD Removal). 4. Removes some nutrients from wastewaters	1. Large land area required. 2. Algae in effluent. 3. Possible septicity and odors. 4. Weed growth, mosquito, and insect problems. 5. Large area requirements.
d. Trickling Filter	Biologically treatable organic wastes	1. Moderate quality effluent (80-85% BOD removal.) 2. Moderate operating costs (lower than activated sludge and higher than oxidation pond). 3. Withstands shock loads better than other biological systems.	1. High capital costs. 2. Clogging of distributors or beds. 3. Snail, mosquito, and insect problems.
e. Chemical Oxidation	Low flow, high concentration wastes of known and consistent waste composition, or removal of refractory compounds.	1. Disinfects effluents. 2. Aids grease removal. 3. Removes taste and odor producing compounds. 4. Removes organics without producing a residual waste	1. Chemical Costs. 2. High initial equipment costs. 3. Skilled operation. 4. Requires handling of hazardous chemicals.

39

(continued)

Table 11. (continued)

f. Spray Irrigation	Non-toxic, dissolved wastes.	concentrate. 1. Inexpensive initial cost. 2. Minimum operator attention.	1. Pollutional removal undefined. 2. Large land area required. High ground porosity needed. 3. Possible contamination of potable aquifers. 4. Freezing in winter. 5. Odors in simmer.
g. Chemical Mixing Flocculation and Clarification	Waste stream high in dissolved solids, colloids, metals, or precipitable inorganics and waste containing emulsified oils.	1. Removes metallic ions, nutrients, colloids, dissolved salts. 2. Recovery of valuable materials. 3. Provides proper conditions for biological treatment.	1. Sophisticated equipment and instrumentation. 2. Residual salts in effluent. 3. Produces considerable sludge.
h. Gravity Filtration	Waste streams with organic or inorganic suspended solids, emulsions, colloids.	1. Breaks emulsions. 2. Removes suspended solids.	1. Clogging. 2. Frequent backwashing.
i. Pressure Filtration	Waste streams high in suspended solids (i.e. sludges, organic solids)	1. High solids removal (95-80%).	1. High equipment costs. 2. Clogging. 3. High pressure drop (power costs).
j. Dissolved-Air Flotation with Chemicals	Waste streams containing oils, fats, colloids, and chemically coalesced materials.	1. Produces high degree of treatment. 2. Removes oils, greases colloids.	1. High initial equipment costs. 2. High operation cost. 3. Sophisticated instrumentation.
k. Anaerobic Contact	Waste streams with high BOD and/or high temperature	1. Methane recovery 2. Small area required. 3. Volatile solids destruction.	1. Heat required. 2. Effluent in reduced chemical form, requires further treatment. 3. Sludge disposal. 4. Requires skilled operation.
4. TERTIARY TREATMENT			
a. Foam Fractionation	Surface active agents (waste streams with high surface	1. Reduces refractory organics. 2. Reduces foaming tendencies	1. Operational costs (Air). 2. Only moderate removal

	tension materials, i.e., anionic ABS)	of effluent.	(30-40% of total organics in secondary effluent).
b. Activated Carbon	Waste streams containing trace amounts of organics and color-, taste, and odor-producing compounds.	1. Removes non-biodegradable organics from wastewaters. 2. Removes taste and odor producing compounds. 3. Reduces color.	1. High equipment costs. 2. Carbon costs- a. pH adjustment b. Initial carbon c. Make-up carbon 3. No inorganic removal. 4. Wastes must be solid-free to prevent clogging. 5. Air pollution potential when regenerating activated carbon. 6. Limited throughout.
c. Stripping	Waste streams containing dissolved gases, volatile organics and materials that can be chemically converted to gases.	1. Removes gaseous or volatile substances. 2. Reduces odors. 3. Chemical recovery. (i.e. sulfur)	1. High initial equipment costs. 2. Fuel costs. 3. Requires pH adjustments. 4. Possible air pollution.
d. Evaporation 1. Solar	Dissolved salts in concentrated solutions, as well as general wastewaters.	1. Low initial cost. 2. Inexpensive operation. 3. Waste volume reduction.	1. Large land area. 2. Dependent on geographical location for evaporation. 3. Solids disposal.
2. Artificial	Dissolved concentrated salts.	1. Waste volume reduction. 2. By-product recovery.	1. Equipment costs. 2. Corrosion and scaling. 3. Heat requirements.
e. Distillation	1. Concentration of inorganic and organ-materials. 2. Separation of inorganic and materials from product water.	1. By-product recovery. 2. Incineration of concentrates. 3. Extremely high effluent quality (≤ 5 mg/L total dissolved solids);	1. Equipment costs. 2. Skilled labor. 3. Energy costs. 4. Maintenance costs (corrosion)

(continued)

Table 11. *(continued)*

		non-selective.	5. Possible air pollution. 6. Extremely high capital and operating costs; volatile matter and ammonia carryover at normal pH; complex equipment and operation; requires pretreatment for removal of suspended solids.
f. Ion Exchange	1. Demineralization 2. Concentration of ionized organics or inorganics.	1. Inorganic removal. 2. By-product recovery. 3. Compact equipment. 4. Selectively concentrates organics and inorganic material. 5. Substitutes less objectionable ions for contaminants. 6. Reliable with equipment easily available 7. Produces effluent with $<$30 mg L total dissolved solids.	1. High equipment costs. 2. Skilled labor. 3. Chemical costs. 4. Organic fouling, requires pretreatment for removal of suspended solids. 5. Disposal of regenerate wastes (brine). 6. Subject to changes in raw waste composition 7. Requires auxiliary chemicals for regeneration.
g. Freezing	1. Sea or brackish waters 2. Dissolved inorganics.	1. Salt separation. 2. Possible by-product recovery	1. High equipment costs. 2. High power costs. 3. Affected by traces of detergents.
h. Electrodialysis	1. Brackish waters. 2. Dissolved inorganics in 750-7500 mg L concentrations.	1. Removes inorganics and some organics. 2. By product recovery. 3. Compact Equipment. 4. Reudces dissolved solids (40-50%).	1. High initial costs. 2. Skilled labor. 3. High power costs. 4. Fouling a. Requires removal of suspended solids and

			dissolved organic solids.
			b. Membrane cleaning and replacement.
		5. Proven plant scale performance.	5. Sophisticated equipment and instrumentation.
			6. Produces excess brine.
i. Reverse Osmosis	Dissolved organics and inorganics.	1. Selectively concentrates inorganics or organics.	1. High initial equipment costs.
		2. Compact Equipment.	
		3. By-product recovery.	2. Very high pressure equipment.
		4. Produces effluent having ≤20 mg L total dissolved solids.	3. Power costs.
			4. Sophisticated instrumentation.
			5. Limited throughput.
			6. Unproven plant scale performance.
			7. Requires extensive pretreatment.
			8. Produces excess brine.
j. Solvent Extraction	Separation of organic materials from water.	1. Selectively concentrates, inorganics and orgnaics.	1. Equipment costs.
		2. Compact equipment.	2. Skilled labor.
		3. By-product recovery.	3. Chemical costs.
			4. Solvent loss.
			5. Solvent recovery.
			6. Limited to low volume wastes.
k. Electro Chemical Degradation	Cyanices and other materials affected by electrochemical reactions.	1. Removes inorganics and organics.	1. Equipment costs.
		2. Compact equipment.	2. Skilled labor.
		3. Destroys bacteria.	3. Energy costs.
			4. Solids disposal.
			5. Air pollution.
			6. Hazardous operation.
l. Subsurface Disposal	Solids-free, concentrated waste streams.	1. Disposal of inorganics and organics.	1. Subsurface clogging.

(continued)

44

Table 11. (continued)

		2. Ultimate disposal of toxic or odorous materials.	2. Ground Water Pollution.
			3. High maintenance and operation costs
			4. Limited Aquifer life.
			5. High initial costs.
m. Ground Water Recharge	Treated waste streams.	1. Reduces bacterial concentration.	1. Possible ground water contamination.
		2. Conserves water resources.	2. Limited to porous formation.
		3. Prevents salt water intrusion into potable aquifers.	
5. SLUDGE (Pretreatment)			
a. Anaerobic Digestion	Biodegradable solids.	1. Methane production.	1. Heat required.
		2. Solids stabilization and conditioning.	2. Process upsets when excess volatile acids generated.
		3. Liquifaction of solids.	3. Odors.
		4. Minimum land required.	4. Skilled labor.
		5. Use of digested sludge as fertilizer or soil conditioner.	5. Explosion hazard.
b. Aerobic Digestion	Biological solids	1. Relatively little odor.	1. Moderate land area required.
		2. Solids stabilization and conditioning	2. Solids disposal required.
		3. Unsophisticated operation.	
c. Autoclaving	Biological solids.	1. Compact operation.	1. High initial equipment costs.
		2. Solids conditioning.	2. Power Costs.
		3. Kills microorganisms.	3. Skilled labor.
			4. Maintenance costs.
d. Elutriation	Sludges with high mineral content or high alkalinity	1. Enhances solids conditioning.	1. Large volumes of water of low alkalinity required.
		2. Chemical savings.	
5. SLUDGE (concentration)			
A. Vacuum Filtration	Organic or inorganic sludges	1. Solids concentration.	1. High equipment, energy, and maintenance costs.

			2. Skilled labor. 3. Necessity for pretreatment (Thickening and chemical addition). 4. Limited throughout.
b. Centrifugation	Non-abrasive, non-corrosive sludges	1. Solids concentration. 2. Compact equipment. 3. Low chemical conditioning. 4. High throughput.	1. Equipment costs. 2. Skilled labor. 3. Energy costs. 4. Maintenance costs.
c. Sand Beds		1. Solids concentration. 2. No chemical costs.	1. Land area required. 2. Weather problems - a. Winter-freezing b. Summer-odor.
d. Presses		3. Low capital costs. 1. Solids concentration. 2. Compact equipment.	
5. SLUDGE (Disposal) a. incineration (regular and fluidized)	Combustible organic sludges (25-33% solids)	1. Excellent sludge volume reduction. 2. Kills biological organisms. 3. Possible by-product recovery a. Heat b. Valuable metals	1. High equipment costs. 2. Fuel and power costs. 3. Air pollution potential. 4. Ash disposal required. 5. Sophisticated equipment and instrumentation.
b. Wet Oxidation	Combustible organic sludges (3-10% solids)	1. Produces easily handled product. 2. Kills biological organisms. 3. Possible by-product recovery. 4. Conditioning prior to other disposal techniques.	1. High initial cost. 2. Fuel and power costs. 3. High organic concentration in effluent stream.

(continued)

Table 11. (continued)

c. Land Disposal	Stable biological sludge	1. Low investment. 2. Postpones ultimate sludge disposal. 3. Provides ultimate disposal, if land is available.	1. Large land area required. 2. Possible odor problem.
d. Sanitary Landfill	Dewatered biological sludges (30-35% solids)	1. Low investment. 2. Suitable for undigested sludges, odorous or toxic materials. 3. Land reclamation.	1. Ground water contamination 2. Requires cover material and compaction. 3. Hauling Costs.

CHAPTER 3

MAINTENANCE MANAGEMENT

Importance of Effective Operation and Maintenance

There are several reasons which demand efficient operation and maintenance in the wastewater treatment plant. First, lack of proper operation and maintenance can result in the failure of the treatment process. The biological process is particularly sensitive to variation from the recommended operational procedure. A detailed description of this process is given along with the recommended operational procedure.

A second reason requiring proper operation and maintenance in the plant is the need to avoid damage to the expensive machinery and equipment. By adhering to the recommended operation and maintenance practices described in this manual, the operator will obtain the longest service life from the facility.

Finally, routine testing and keeping current and complete records will make it possible to detect and correct minor problems before these problems become more severe.

Operational and Managerial Responsibility

The efficient and economical operation of the municipal wastewater treatment facility is the responsibility of both the operational personnel and the management owning the facility. The objectives of the operators and the municipality should be to run the treatment plant in an economical manner. The plant operator will report directly to the Director of Water and Sewers and the director in turn will report to the government agency (City Council, etc.). To achieve this objective, the responsibility of the operators and management are defined as follows:

1. *Operator's Responsibilities*

   a. The operator will be familiar with the treatment plant, its processes and their functions; and will operate the plant effectively.

   b. He will keep himself informed of the current operating and maintenance practices of the plant he is responsible to operate.

   c. He will regularly read the periodicals supplied by the municipality to enhance his knowledge of the treatment facility, and will participate in short courses and schools when available.

   d. He will maintain accurate operational and maintenance records of the treatment facility.

   e. He will use sound judgment in the expenditure of operating funds.

f. He will inform the municipality of potential major problems in operation and maintenance of the system.

g. He will assist supervisors in preparing an adequate budget.

h. He will be aware of the safety hazards in the sewage treatment plant.

i. He will be prepared to discuss plant operation with superivsors and plant visitors and will maintain good public relations.

j. He will know expected efficiencies of treatment units and how to monitor these units to achieve expected efficiencies.

2. *Manager's Responsibilities*

a. Provide good working conditions, safety equipment and tools for operational personnel.

b. Provide operator training program.

c. Make periodic inspections of the treatment system to discuss mutual problems with operation personnel.

d. Maintain good public relations.

e. Prepare budgets and reports.

f. Plan for future needs of the facility.

General Maintenance

For the effective operation of any treatment facility, periodic and preventive maintenance is essential. Without maintenance, the mechanical equipment will not continue to function properly and will cause deterioration in plant performance due to inadequate chemical, physical or biological operations. Preventive maintenance will minimize equipment failure. Periodic maintenance of all facilities will keep the plant running efficiently.

Mechanical maintenance is of prime importance as the equipment must be kept in good operating condition in order for the plant to maintain peak performance, Manufacturers provide information on the mechanical maintenance of their equipment. You should thoroughly read their literature on your plant equipment and understand the procedures. Contact the manufacturer or his local representative if you have any questions. Follow the instructions very carefully when performing maintenance on equipment. You also must recognize tasks that may be beyond your capabilities or repair facilities, and you should request assistance when needed.

For a successful maintenance program, your supervisors must understand the need for and benefits from equipment that operates continuously as intended. Disabled or improperly working equipment is a threat to the quality of the plant effluent, and repair costs for poorly maintained equipment usually exceed the cost of maintenance.

Maintenance requires considerable skill which can only be acquired by experience, study and practice. Basically, any maintenance program should start with good housekeeping and should observe the following:

1. Keep a clean, neat and orderly plant.
2. Adhere to a systematic plan for the execution of daily operations.
3. Adhere to a routine schedule for inspection and lubrication.
4. Keep data and records of each piece of equipment, listing unusual incidents and faulty operating conditions. From a review of records, an operator can determine the weaknesses of various pieces of equipment and will be able to determine which spare parts should be kept in stock.
5. Observe good safety practices.

Preventive Maintenance Summary Schedule

A schedule of routine preventive maintenance follows as a convenience to the operator. If any information in the schedule differs from a manufacturer's recommendations, observe the recommendation of the manufacturer.

The frequency of performing some items of maintenance will depend upon the actual operating and environmental plant conditions. After the operator gains experience with the various pieces of equipment, he may be able to vary the frequency of some of the maintenance procedures from these recommended frequencies. If the operator finds that additional items of maintenance are required, he should add them to the maintenance schedule.

Every Day

- Check motors for overheating and excessive vibration.

Every Week

- Check oil level in comminutor and add oil, if necessary.
- Check oil level in comminutor extension bearing and add oil, if necessary.
- Check grease at lift station sewage pump bearings and add grease, if necessary.
- Operate engine in lift station for at least 30 minutes.
- Check grease at grit washer conveyor screw drive end shaft and add grease, if necessary.
- Check oil level in grit washer conveyor screw worm gear reducer and add oil, if necessary.
- Check blower gearbox oil level and add oil, if necessary.
- Check blower bearing oil levels and add oil, if necessary.
- Check air diffusers for clogging by visually noting water surface above diffuser assemblies and clean, if necessary.
- Check oil level in gear and top bearing compartment of flash mixing tank mixer and add oil, if necessary.
- Check grease at flash mixing tank mixer lower bearing and add grease, if necessary.
- Check oil level in clarifier mechanism worm gear reducer and add oil, if necessary.
- Check for condensation in clarifier mechanism worm gear reducer and drain, if necessary.

- Check lubrication of return sludge pump bearings and add grease, if necessary.
- Check grease at re-use water pump bearings and add grease, if necessary.

Every 2 Weeks

- Lubricate lift station sewage pump drive shafts.
- Check the lift station sewage pump adjustable speed drive for accumulation of dirt on the power modules, diode module heat-sink pins, and printed circuit boards inside the control circuit enclosure and clean, if necessary. Inspection of printed circuit board must be done with *all A.C. power removed.*
- Check lift station engine battery connections and battery electrolyte level.
- Wipe any residue from flow recorder pin points.

Every Month

- Inspect lift station sewage pumps for proper lubrication, leaking seals, and proper performance and lubricate, if necessary.
- Replace strip charts on both flow recorders.
- Check flow recorder ink levels and add ink, if necessary.
- Check belt tension on grit washer conveyor screw drive and adjust, if necessary.
- Check air blower air filters and clean or replace, if necessary.
- Check grease at air diffuser ease-out elbows and add grease, if necessary.
- Clean chlorinator water and chlorine strainers on both chlorinators.
- Inspect the flow meters and rate valves of both chlorinators and clean, if necessary.
- Inspect the chlorine distribution panel varea-meters and clean, if necessary.
- Check the packing in the plug valves for leaks and tighten the gland followers, if necessary.

Every 3 Months

- Inspect the comminutor cutter bars for sharpness and to see that a rubbing fit exists between the oscillating and stationary cutters and sharpen or adjust, if necessary.
- Change the oil in the grit washer conveyor screw worm gear reducer.
- Change blower gearbox oil.
- Change blower bearing oil.
- Drain and replace oil in the gear and top bearing compartment of the flash mixing tank mixer.
- Change the clarifier mechanism gear motor oil.

Every 6 Months

- Change the comminutor gearbox oil.
- Check the lift station sewage pump wear plates and wear rings and replace, if necessary.

50

- Lubricate the casters of the air diffuser assembly hoist.
- Change the oil in the air diffuser.
- Change the clarifier mechanism worm gear reducer oil.

Every Year

- Lubricate comminutor upper motor bearing.
- Lubricate comminutor intermediate shafting.
- Clean the vacuum and pressure regulating valves on both chlorinators.
- Clean both chlorinator injectors.

Special

- Change the comminutor oil whenever the comminutor requires dismantling.
- After the first week of operation, drain the oil from the comminutor gearbox and replace with light SAE-10 oil. Run for three minutes and drain. Refill with a type of oil recommended by the manufacturer.
- Check the oil level in the air diffuser assembly hoise winch before every use and add oil, if necessary.
- Change the clarifier mechanism gear motor oil after the first 200 hours of operation.
- Lubricate the plug valves whenever they are dismantled.
- For the recommended schedule of maintenance for the various motors, see the manufacturers' recommendations delivered with the motors.

General Housekeeping Schedule

In order to keep the treatment plant attractive, to prevent nuisance conditions from arising, and to prevent a progressively deteriorating plant, general housekeeping chores must be completed on a regular basis.

A list of the minimum recommended general housekeeping chores follows. If the operator finds the additional items are required, he should add them to the list.

1. General year maintenance, including lawn cutting, lawn watering, and litter removal.
2. Janitorial services in the Control House.
3. Janitorial services in the Equipment Building.
4. Cleaning grease, scum or solids from the side walls, weirs, and other treatment tank surfaces.
5. General building maintenance, including paint touch-up and structural repairs.
6. Clean bar screens and dispose of the screenings.
7. Dispose of collected grit.
8. Clean up spillages and stoppages.
9. Inspect coatings of steel structures and paint areas where the coating is damaged.

The frequency of performing some of the listed items will depend upon the actual operating and environmental conditions at the plant. After gaining experience with the plant, the operator will be able to determine the required frequency for each item. It is recommended that initially the general housekeeping chores be completed at least twice a week.

The Maintenance Management System [1]

The following is a brief introduction to each of the five basic features of any sound maintenance management system. A system efficiently using these basic features will provide for economy in maintenance cost and reliability in equipment operations for the maintenance program.

Equipment Record System — The equipment record system contains information on each item of equipment requiring maintenance. This system may be one card on each item of equipment, a number of cards for each item or for large plants, a combination of both cards and data maintained on a computer.

The equipment record system provides information on preventive maintenance tasks with their frequencies, corrective maintenance work performed and maintenance cost data. Information used in making cost analyses, preparing maintenance budgets and evaluating maintenance problems will be found in the record system.

Maintenance Planning and Scheduling — The planning and scheduling of maintenance work is an essential ingredient in any good maintenance management system. If planning and scheduling is done properly, the maintenance force will be used in the most efficient manner and the preventive maintenance program will be effective. Tools for planning and scheduling include schedule boards, maintenance work orders, and maintenance labor standards. The key to successful planning and scheduling is a realistic estimate of a facility's corrective and preventive maintenance needs.

Storeroom and Inventory System — The purpose of a good maintenance management system is to ensure the proper operation of the treatment facility. The storeroom and inventory procedures used at the treatment plant can help ensure that the maintenance system's purpose is achieved. A good storeroom procedure will maintain control of items on hand, will recognize when to reorder needed supplies, will facilitate locating items on hand, and will provide for efficient purchasing and receiving of all supplies.

Maintenance Personnel and Organization — Regardless of the care which goes into the selection of an equipment record system or work order system it is the plant's maintenance staff who is ultimately responsible for ensuring the maintenance management system functions properly. All maintenance tasks within the treatment facility should be analyzed, and properly trained personnel in sufficient numbers should be provided. The maintenance management system should provide adequate staffing, arrange comprehensive maintenance training programs, develop accurate job descriptions for all personnel, and maintain an organizational chart to help define maintenance responsibilities.

Cost and Budgets for Maintenance Operations — Information on plant maintenance cost and the development of a maintenance budget are very important for their incorporation into the plant's total operation and maintenance budget. Before an accurate estimate of maintenance cost can be made or a sound maintenance budget can be prepared, it is necessary to divide the maintenance operations into service categories such as preventive maintenance, corrective maintenance, and major repairs or alterations. With the maintenance operations defined, the information in the equipment card file on work performed, work contracted out, items used from storeroom stock and purchased, and a breakdown of man-hours can be used to develop information on maintenance cost. With allowances for equipment replacement, expansion, and information on maintenance history for the plant, the maintenance budget then can be developed.

EQUIPMENT RECORD SYSTEM

An equipment record system should be developed to maintain information on each item of equipment. The facility owner should be involved in selecting the system to be recommended. The equipment record system should contain information on each item of equipment. This system may be only one card on each type of equipment, a number of cards for each item or for larger plants, a combination of information cards and data maintained on a computer.

The following is an outline of the information the equipment cards or computer should contain:

- Description of equipment and equipment number with location in plant.
- Supplier with address, representative, phone number, date of purchase with cost.
- Size, model, type, and serial number.
- Electrical and/or mechanical data.
- Inventory of spare parts on hand.
- Preventive maintenance items to be accomplished with their frequency. Space to note when PM was performed, by whom and pertinent comments. Data on man-hours, costs, and materials or supplies consumed.
- Information on corrective maintenance work should be maintained in a manner similar to that outlined above for preventive maintenance.
- A system to compile this information for use in determining cost and for future use in budget development.

Adequate maintenance records assist in equipment evaluation, aid in establishing budgets for manpower and materials, and help ensure an efficient schedule for preventive maintenance functions. Maintenance records also provide plant management with an indication of the efficiency of the maintenance personnel. This allows management to assign maintenance personnel to jobs for which they are best qualified and to rotate personnel for training and overall experience throughout the facility. Such experience provides for continuity during absences or personnel changes. The information in the equipment record system can help provide answers to the following maintenance related questions:

1. Is the maintenance work on a specific item of equipment excessive in relation to
   other similar units?
   a. What is the cause?
      1. Is it lack of preventive maintenance?
      2. Is it wrong application of the equipment?
      3. Is breakdown result of inherent defect in the equipment?
      4. Is it due to poor lubrication?
      5. Was the equipment overloaded?
2. Is this maintenance procedure repetitive?
   a. Can the maintenance procedure be simplified, improved or eliminated?
   b. Are the correct tools available?
   c. Was an experienced craftsman assigned to the job?
   d. Were spare parts required?
      1. Were the spare parts immediately available from plant inventory?
   e. Can preventive maintenance procedures be established at less cost?
3. Was overtime work required?
4. Is the clerical time required out of proportion to the maintenance work performed?
   a. Is the overhead cost of clerical work justified for this facility?
5. Can the maintenance be done by contract with outside party?
6. Do the maintenance records help in planning maintenance work and scheduling?
   a. Can the cost of time and materials be reduced?
7. Is the data of sufficient volume for the facilities maintenance functions to be
   processed by computer?

Maintenance Planning and Scheduling

A work order system should be used to initiate all preventive maintenance and
corrective maintenance tasks. There are two basic types of work orders. The first
type is a standard work order and generally covers repetitive work to be done in a
given area, such as a preventive maintenance task. The second type work orders are
written job orders. They may origniate in the area where the work is to be performed and constitute the authority to carry out specific corrective maintenance tasks.
They can also be originated by the maintenance supervisor to perform corrective
maintenance tasks identified either by verbal reports or conditions noted where
performing preventive maintenance tasks. The following are the basic features a
good work order system should contain:

1. The work order form should contain the following minimum information:
   date; work order number; location of work to be performed; nature of problem; the work required; the desired timing for accomplishing the job, such as
   emergency, as soon as possible, when convenient, or during equipment shutdown; space to write information on actual work accomplished and comments;
   and space for supervisory signature.

2. The work order becomes a record of repairs and a history of the equipment requiring repair. It also provides a method for comparing similar items of equipment with respect to their maintenance requirements.
3. The work order establishes a comparative guide for the cost of repetitive repairs; the cost of similar jobs can be checked for cost overruns.
4. The work order should not be financed from petty cash but included in budget considerations.
5. The work order should be a guide for cost and materias expanded.
6. The work order should define the work to be performed, the materials required, and the schedule to be followed for a particular job.
7. Only special emergency work should be performed without first preparing a work order. For emergency repairs, a work order should be filed to complete the maintenance records.
8. The cause of the needed repair should be reported on the work order.
9. The work order should be a complete record of the repair service.
10. The work order should be signed only by authorized persons.
11. The work order system should help to reduce equipment downtime.
12. The work order records whether the repair could have been avoided with adequate preventive maintenance.
13. The work order should be used whenever a repair requires new parts, equipment shutdown, or outside repair service. Routine maintenance below a pre-established time limit does not have to be placed on work orders.
14. With the work completed, the work order should be used to note the task on the equipment record for the particular item of equipment. The task description, work order number, and the cost data with date should be noted in the record system. The work order number will provide access to locating the work order should the need arise.
15. The work orders should be filed by work order number and kept as a history record of work performed.

BIBLIOGRAPHY

1. *Maintenance Management Systems for Municipal Wastewater Facilities*, Municipal Operations Branch, Office of Water Programs Operations, USEPA, Washington DC, Contract No. 68-01-0341, October 1973.

CHAPTER 4

MONITORING

INTRODUCTION

EPA has expressed the need for monitoring in its outline of objectives in the Federal Register. [1]

"The objectives of the State monitoring program required by the Act and these regulations are to determine compliance with permit terms and conditions, to develop and maintain an understanding of the quality (and causes and effects of such quality) of the waters in the State for the purpose of supporting State water pollution control activities, to report on such quality and its causes and effects, and to assess the effectiveness of the State's pollution control program. To this end each State shall carry out a broad range of monitoring activities both before and after implementing pollution controls, including measurement of pollutant sources, water quality, the factors affecting water quality, and the specific effects of such quality upon beneficial uses of the State's waters."

Most states will pursue this activity through the monthly operating reports received from sewage treatment plant operators. In addition to the State requirements, plants will be interested in monitoring to:

1. Evaluate plant efficiency,
2. Evaluate individual unit efficiency,
3. Protect the plant from toxic materials or surges of industrial wastes.

Requirements and Purposes of Sampling [2]

"Sampling of sewage is performed to satisfy various purposes and requirements. These include the planning, design and operation of facilities for the control and treatment of sewage; the enforcement of water quality standards and objectives; and general research to increase our knowledge of the characterization of sewage."

"Development of a program of sampling is presently based on a limited number of properties and constituents for which analyses are made. The type of sample collected depends on the purpose of the program, and on both technical and economic considerations."

Common Properties and Constituents [2]

"Although the constituents of sewage include most substances known to man, there are a limited number of analyses made to determine the more common properties and constituents. Columns of the following table provide information as follows:

Column 1 — Specific sewage parameters.

Column 2 — Preservation essential to sustaining the character of sewage as sampled between the time of sampling and the time of making the analyses.

Column 3 — The maximum period for holding the sample prior to the analysis. This is the allowable period after the preservative has been applied.

Column 4 — Approximate required sample size. This can be considered only a very rough figure to assist in designing a sampling program. In some cases there may be a difference in the size of sample required for a given parameter, depending on the method of analysis to be used."

"Figures given for sample size are generally large. For example, much smaller samples are needed with use of various systems of automatic analysis. The Technicon Auto-Analyzer requires samples of less than 30 ml, and is recommended for total alkalinity, chloride, cyanide, fluoride, total hardness, nitrogen (ammonia), nitrogen (Kjeldahl), nitrogen (nitrate — nitrite), phosphorus, sulfate, COD, and others."

Type of Sample [2]

"The type of sample collected depends on a number of factors such as the rate of change of flow and of the character of the sewage, the accuracy required, and the availability of funds for conducting the sampling program. All samples collected, either manually or with automatic equipment, are included in the following types:

1. Manual "grab" samples which are obtained by dipping a container into the sewer and bringing up a sample of wastewater. Containers are sometimes devised to grab a sample at a stationary depth or so that a sample integrated from bottom to top of the stream is collected. Water flows gradually into the container as it passed through the flow.

2. Automatic "grab", or discrete, samples which are collected at selected intervals, and each sample is retained separately for analysis. Usually each sample is collected at a single point in the sewer cross-section. However, in a few instances samplers with multiple ports have been used to allow simultaneous collection from several points in the cross-section.

3. Simple composite samples, which are made up of a series of smaller samples (aliquots) of constant volume collected at regular time intervals and combined in a single container. The series of samples is collected over a selected time period, such as 24 hours, or during a period of storm runoff, for example. The simple composite represents the average condition of the

Table 1. Properties and Constituents of Sewage.

Parameter	Preservative	Maximum Holding Period	Approximate Required Sample Size (ml)
Acidity-Alkalinity	Refrigeration at 4°C	24 hours	50-100
Biochemical Oxygen Demand (5-day)	Refrigeration at 4°C	6 hours	1000
Calcium	None Required	Indefinite	50
Chemical Oxygen Demand	2 ml H_2SO_4 per liter	7 days	50
Chloride	None Required	Indefinite	50
Bacteria - fecal coliform, total coliform, or fecal streptococcus	Maintain temperature as at source - usually requires refrigeration	8 hours	200
Color	Refrigeration at 4°C	24 hours	50
Cyanide	NaOH to pH 10	24 hours	500
Dissolved Oxygen	Determine on site	No holding	250-300
Fluoride	None Required	7 days	200-300
Hardness	None Required	7 days	25
*Metals, Total	5 ml HNO_3 per liter	6 months	100
**Metals, Dissolved	Filtrate: 3 ml 1:1 HNO_3 per liter	6 months	100
Nitrogen, Ammonia	40 mg $HgCl_2$ per liter - 4°C	7 days	100
Nitrogen, Kjeldahl	40 mg $HgCl_2$ per liter - 4°C	Unstable	100
Nitrogen, Nitrate - Nitrite	40 mg $HgCl_2$ per liter - 4°C	7 days	100
Oil and Grease	2 ml H_2SO_4 per liter - 4°C	24 hours	1000
Organic Carbon (Total and Dissolved)	2 ml H_2SO_4 per liter (pH 2)	7 days	100
pH	Determine on site	No holding	--
Phenolics	1.0 g $CuSO_4/1$ + H_3PO_4 to pH 4.0 - 4°C	24 hours	500
Phosphorus	40 mg $HgCl_2$ per liter - 4°C	7 days	200
Solids (total, dissolved, suspended, volatile)	None Available	7 days	1000
Specific Conductance	None Required	7 days	--
Sulfate	Refrigeration at 4°C	7 days	100
Sulfide	2 ml Zn acetate per liter	7 days	1000
Threshold Odor	Refrigeration at 4°C	24 hours	200
Turbidity	None Available	7 days	1000

* Sum of the concentrations of metals in both the dissolved and suspended fractions.

** For determination of trace metals in solution by atomic absorption spectroscopy.

waste during the period only if the flow is constant.

4. Flow-proportional composite samples, which are collected in relation to the flow volume during the period of compositing, thus indicating the "average" waste condition during the period. One of two ways of accomplishing this is to collect samples of equal volume, but at time intervals which are inversely proportional to the volume of flow. That is, the time interval between samples is reduced as the volume of flow increases, and a greater total sample volume is collected. Flow proportioning can also be achieved by increasing the volume of each sample collected in proportion to the flow, but keeping the time interval between samples constant.

5. Manually composited samples which are obtained, where recording flow records are available, from fixed volume "grab", or discrete, samples collected at known times and proportioned manually to produce a flow proportioned composite sample.

6. Sequential composite samples, which are composed of a series of short-period composites, each of which is held in an individual container. For example, each of several samples collected during a 1-hour period may be composited for the hour. The 24-hour sequential composite is made up from the individual 1-hour composites."

Adequacy of a Sampling Program [2]

"The adequacy of a sampling program depends largely on the optimum selection of sampling sites. Both the program cost and its effectiveness in collecting samples representative of the character of sewer flows are seriously affected by the care exercised in site selection. Similarly, the kinds of samplers selected determine the adequacy of the program with respect to obtaining suitable data for the needs of the particular sampling program.

"In most cases, use of mathematical statistical analysis for determining the probable errors in the data obtained by sewer sampling is not practical. A single "grab" sample of 1 liter, even in dry-weather flows, is not necessarily indicative of the average character of the flow. With respect to an instant of time, the indicated character of the sewage may vary with the point in the cross-section from which it was "grabbed". One must consider the universe of sewage volumes represented by the sample. At the instant of sampling, it may be all the liters of sewage in the cross-section at that instant. But, if the sewage is not thoroughly mixed, we know that the sample is biased, that is, it may represent only a portion of the 1-liter samples in the cross-section, possibly only those near the surface of the flow.

"In periods of storm runoff, it is known, if only by observation, that the character of the sewage is continually changing, possibly with great rapidity. There then becomes no single universe represented by the "grab" sample. Instead, there is an infinite number of universes, and the single "grab" sample is without meaning in determining the character of the sewage. A similar situation exists in the case of sewers carrying industrial wastes. The variability of low and of quality parameters during periods of storm runoff are illustrated in Figure 1, wherein quantity and quality data for a storm on the Bloody Run sewer watershed at Cincinnati, Ohio, are graphically presented.

"It becomes apparent, then, that a large number of samples is required to adequately characterize the character of sewage in a combined sewer during and immediately after a storm event, particularly if the character is to be related to flow rate. Compositing the samples in proportion to flow rate may determine the average character of the sewage during the period of compositing. However, it does nothing to describe the pattern of changes which may occur during that period.

"Awareness of the general character of sewer flows and of flow modes in storm sewers and combined sewers, and knowledge of the variability of pollutant concentration, leads to an understanding of how best to select sites for sampling. Some of the considerations in making such selections are:

1. *Maximum accessibility and safety* — Manholes on busy streets should be avoided if possible; shallow depths with manhole steps in good condition are desirable. Sites with a history of surcharging and/or submergence by surface water should be avoided if possible. Avoid locations which may tend to invite vandalism.

2. *Be sure that the site provides the information desired* — Familiarity with the sewer system is necessary. Knowledge of the existence of inflow or outflow between the sampling point and point of data use is essential.

3. *Make certain the site is far enough downstream from tributary inflow* to ensure mixing of the tributary with the main sewer.

4. *Locate in a straight length of sewer,* at least six sewer widths below bends.

5. *Locate at a point of maximum turbulence,* as found in sewer sections of greater roughness and of probable higher velocities. Locate just downstream from a drop or hydraulic jump, if possible.

6. *In all cases, consider the cost of installation,* balancing cost against effectiveness in providing the data needed.

"Presently available sewage samplers have a great variety of characteristics with respect to size of sample collected, lift capability, type of sample collected (discrete or composite), material of construction, and numerousother both good and poor features. A number of considerations in selection of a sampler are:

1. Rate of change of sewage conditions.
2. Frequency of change of sewage conditions
3. Range of sewage conditions
4. Periodicity or randomness of change
5. Availability of recorded flow data
6. Need for determining instantaneous conditions, average conditions, or both
7. Volume of sample required
8. Need for preservation of sample
9. Estimated size of suspended matter

10. Need for automatic controls for starting and stopping
11. Need for mobility or for a permanent installation
12. Operating head requirements.

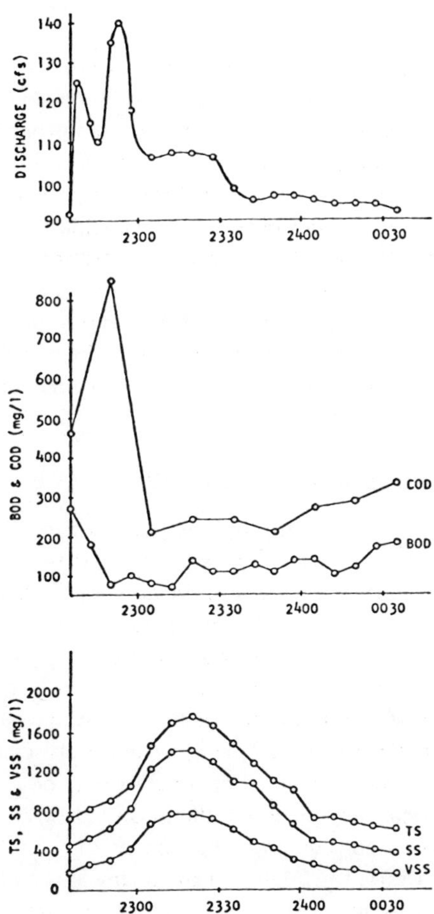

*Figure 1. Runoff Quantity and Quality Data,
Bloody Run Sewer Watershed.*

"Because of the variability in the character of storm and/or combined sewage, and because of the many physical difficulties in collecting samples to characterize the sewage, precise characterization is not practicable, nor is it possible. In recognition of this fact, one must guard against embarking on an excessively detailed sampling program, thus increasing costs, both for sampling and for analyzing the samples, beyond costs that can be considered sufficient for conducting a program which is adequate for the intended purpose.

"A careful study of costs should be made prior to commencing a program of sampling, balancing cost against the number of samples and analyses required for adequate characterization of the wastewater. As the program progresses, current study of the results being obtained may make it reasonable to reduce or increase the number of samples collected.

"The unit cost of handling and analyzing samples can often be reduced by careful planning and scheduling of field work, and by coordination with laboratory requirements. If the volume of samples is large, and the program is to continue over a long time period, consideration should be given to use of equipment for automatic analyses and in-situ monitoring. A number of equipment types and methods, such as specific ion electrodes and probes, are available for these purposes. As an example, approximately 15 samples per hour can be analyzed for chloride, using the Technicon Auto-Analyzer. Samples of only 4 ml volume are required. Caution is needed in selecting equipment suitable for a series of parameters for which analyses are to be made. With some equipment, the time required for making necessary adjustments between each of a series of tests may counteract the rapidity of making analyses for a single parameter."

Specific Sampling Purposes and Requirements [2]

"Sampling programs are set up for various purposes for which the requirements are not necessarily the same. That is, parameters important to one kind of project may not be needed for another project having a different objective. As an example, parameters of interest for operation of facilities for control and treatment of stormwater and/or combined sewage may be more limited in number than those needed for planning and design of the facilities. In the operating stage, experience at the particular location and with the unique facilities, may have demonstrated a more limited sampling need. On the other hand, where stormwater is combined with industrial wastes, analyses for additional parameters may be required.

"A number of physical, chemical, combinations of physical-chemical, and biological methods have been considered in the Storm and Combined Sewer Pollution Control Program of the EPA for treatment of stormwater and combined sewage. In most cases, some type of control such as reduction of instantaneous peak flows is essential for practical application of treatment methods. These include storage facilities of many types, flow regulation and routing, and remote flow and overflow sensing and telemetering.

"Specific processes which have been investigated are:
Physical — (1) Fine mesh screening; (2) Microstrainer; (3) Screening/Dissolved-air flotation; (4) High-rate single-, dual, or tri-media filtration; (5) Swirl and helical separation; (6) Tube settlers; etc.
Chemical — (1) Coagulant and polyelectrolyte aids for sedimentation, filtration, flotation and microstraining; (2) Chemical oxidation and use of ozone for oxida-

tion; and (3) Disinfection — chlorination, ozonation, high rate application, on-site generation, and use of combined halogens (chlorine and iodine) and chlorine dioxide.

Physical-Chemical — (1) Screening plus dissolved-air flotation with flotation aids; (2) Screening — chemical flocculation — sedimentation — high rate filtration; (3) Powdered and granular activated carbon adsorption; (4) Chemical flocculation — tube sedimentation — tri-media filtration; and (5) Screening — coagulation — high rate dual-media filtration.

Biological — (1) High-rate plastic and rock media trickling filters; (2) Bio-adsorption (contact stabilization); (3) Stabilization ponds; (4) Rotating biological contactor; and (5) Deep-tank aerobic and anaerobic treatment.

"For planning and designing such facilities and processes, and for testing their impact on receiving streams, sampling for certain basic wastewater parameters is essential. In general these include:

1. *Biochemical oxygen demand (BOD)* — Used to determine the relative oxygen requirement of the wastewater. Data from BOD tests are used for the development of engineering criteria for the design of wastewater treatment plants.

2. *Chemical oxygen demand (COD)* — Provides additional information concerning the oxygen requirement of wastewater. It provides an independent measurement of organic matter in the sample, rather than being a substitute for the BOD test. For combined sewer overflows and stormwater, COD may be more representative of oxygen demand in a receiving stream because of the presence of metals and other toxicants which are relatively non-biodegradable.

3. *Total oxygen demand (TOD)* — A recently developed test to measure the organic content of wastewater in which the organics are converted to stable end products in a platinum-catalyzed combustion chamber. The test can be performed quickly, and results have been correlated with the COD in certain locations.

4. *Total organic carbon (TOC)* — Still another means of measuring the organic matter present in water which has found increasing use in recent times. The test is especially applicable to small concentrations of organic matter.

5. *Chloride* — One of the major anions in water and sewage. The concentration in sewage may be increased by some industrial wastes, by runoff from streets and highways where salt is used to control ice formation, salt water intrusion in tidal areas, etc. A high chloride content is injurious to vehicles and highway structures, and may contaminate water supplies near the highway.

6. *Nitrogen Series* — A product of microbiologic activity, is an indicator of sewage pollution, or pollution resulting from fertilizers, automobile exhausts, or other sources. Its presence may require additional amounts of

chlorine, or introduction of a nitrogen fixation process, in order to produce a free chlorine residual in control of bacteria.

7. *pH* — The logarithm of the reciprocal of hydrogen ion activity. State
regulations often prescribe pH limits for effluents from industrial waste
treatment plants. Provides a control in chemical and biological treatment
processes for wastewater.

8. *Solids (Total, Suspended, Volatile, and Settleable)* — Usually represent a
large fraction of the pollutional load in combined sewage. Inorganic sedi-
ments, in a physical sense, are major pollutants, but also serve as the
transporting or catalytic agents that may either expand or reduce the
severity of other forms of pollution.

9. *Oil and Grease* — Commonly found in sanitary sewage, but also appear in
industrial wastes as a result of various industrial processes. Present a
serious problem of removal in wastewater treatment facilities.

10. *Bacterial Indicators (Total Coliform, Fecal Coliform, Fecal Strepto-
coccus)* — Indicate the level of bacterial contamination.

"Where more exotic wastes are combined with stormwater and sanitary sewage,
additional treatment facilities may be required for the removal of industrial bypro-
ducts and nutrients such as cyanide, fluoride, metals, pesticides, nitrogen, phosphorus, sulfate and sulfide. For planning and design of such treatment facilities,
additional analyses are required in accordance with the pollutant material expected
in the wastewater. This may, in turn, require significant expansion of the sampling
program.

"Sampling and analyses of wastewater are necessary to the satisfactory operation
of treatment plants. Pollutants in the incoming storm sewer or combined sewer are
compared with those in the effluent from the treatment plant to determine the
effectiveness of the treatment process. Additionally, sampling of the receiving
stream before and after treatment/control system installation indicates the benefits
gained from the installation. Knowledge of the concentration of pollutants entering
the plant can be used also to make adjustments to the treatment process as required. Continuous monitoring of the stream below the treatment/control facility is
important to facility operation. Depending on the type or types of treatment
process used, the number of parameters required for sampling and analyses is usually less than those required for planning and design. For example, where treatment
consists only of sedimentation and chlorination, analyses for oxygen demand, suspended solids, bacterial indicators, and for chlorine residual may be sufficient. If
chemicals are used to assist the sedimentation process, determination of pH may be
needed. The sampling program can be determined largely in accordance with previous experience and knowledge of the pollutants found.

"Sampling programs should start long before installation of combined sewer
overflow and stormwater treatment/control facilities to establish the objectives of
the facilities and to provide necessary design and operation criteria. A much longer
time period for sampling may be required than anticipated because of the need to

sample during periods of storm runoff, which may be few in drought years.

"In some cases, the availability of historical quality data may provide a basis for prediction of future character for planning and design purposes. Dependence on such predicted data is not sufficient, and collection of current data is required to verify predictions and, later, to measure facility effectiveness.

"Programs of sampling and analyses of wastewater in storm and/or combined sewers are frequently used for the enforcement of water quality standards or objectives. Such programs provide information leading to the source of various types of pollution. Often, the wastewater is continually monitored to check on compliance with pollution control laws and regulations. The range of different parameters to be measured for these purposes is continually expanding with the development of new processes. There appears to be no limit to future analytical requirements."

DESIRABLE EQUIPMENT CHARACTERISTICS [2]

Equipment Requirements [2]

"The success of an automatic sampler in gathering a representative sample starts with the design of the sampler intake. This obviously will be dependent upon conditions at the particular site where the sample is to be extracted. If one is fortunate enough to have a situation where the sewer flow is homogeneous with respect to the parameters being sampled, then a simple single point of extraction for the sample will be adequate. In the more typical case, however, there is a spatial variation in the concentration of the particular constituent that is to be examined as part of a sampling program, and then the sampling intake must be designed so that the sample which is gathered will be nearly representative of the actual flow. Several different designs have been utilized in an attempt to meet this objective . However, none can be considered as ideal or universally applicable. "The automatic sampler must be capable of lifting the sample to a sufficient height to allow its utilization over a rather wide range of operating heads. It would appear that a minimum sample lift of 15 feet or so is almost mandatory in order to give a fairly wide range of applicability. It is also important that the sample size not be a function of the sample lift; that is, the sample size should not become significantly less as the sample lift increases. "The sample line size must be large enough to give assurances that there will be no plugging or clogging anywhere within the sampling train. However, the line size must also be small enough so that complete transport of suspended solids is assured. Obviously, the velocities in any vertical section of the sampling train must well exceed the settling velocity of the maximum size particle that is to be sampled. Thus, the sample flow rate and line size are connected and must be approached together from design considerations.

"The sample capacity that is designed into the piece of equipment will depend upon the subsequent analyses that the sample is to be subjected to and the volumetric requirements for conducting these analyses. However, in general, it is desirable to have a fairly large quantity of material on hand, it being safer to err on the side of collecting too much rather than too little. As a minimum, it would seem that at least a pint and preferably a liter of fluid would be desirable for any discrete sample. For composite samples at least a gallon and preferably two (4 to 8 liters) should be collected.

"The controls on the automatic sampler should allow some degree of freedom in the operation and utilization of the particular piece of equipment. A built-in timer is desirable to allow preprogrammed operation of the equipment. Such operation would be particularly useful, for example, in characterizing the buildup of pollutants in the early stages of storm runoff. However, the equipment should also be capable of taking signals from some flow measuring device so that flow porportional operation can be realized. It is also desirable that the equipment be able to start up automatically upon signal from some external device that might indicate the onset of storm flow phenomena such as an external rain gauge, flow height gauge, etc. Flexibility in operation is very desirable.

"A power source will be required for any automatic sampler. It may take the form of a battery pack or clock type spring motor that is integral to the sampler itself. It may be pressurized gas, air pressurized from an external source, or electrical power, depending upon the availability at the site.

"In addition to being able to gather a representative sample from the flow, the sampling equipment must also be capable of transporting the sample without pre-contamination or cross-contamination from earlier samples or aliquots and of storing the gathered sample in some suitable way. As was noted in section IV, chemical preservation is required for certain parameters that may be subject to later analyses, but refrigeration of the sample is also required and is stated as the best single means of preservation."

Desirable Features [2]

"In addition to the foregoing requirements of automatic sampling equipment, there are also certain desirable features which will enhance the utility and value of the equipment. For example, the design should be such that maintenance and troubleshooting are relatively simple tasks. Spare parts should be readily available and reasonably priced. The equipment design should be such that the unit has maximum inherent reliability. As a general rule, complexity in design should be avoided even at the sacrifice of a certain degree of flexibility of operation. A reliable unit that gathers a reasonably representative sample most of the time is much more desirable than an extremely sophisticated complex unit that gathers a very representative sample 10 percent of the time, the other 90 percent of the time being spent undergoing some form of repair due to a malfunction associated with its complexity.

"It is also desirable that the cost of the equipment be as low as practical both in terms of acquisition as well as operational and maintenance costs. For example, a piece of equipment that requires 100 man-hours to clean after each 24 hours of operation is very undesirable. It is also desirable that the unit be capable of unattended operation and remaining in a standby condition for extended periods of time.

"The sampler should be of sturdy construction with a minimum of parts exposed to the sewage or to the highly humid, corrosive atmosphere associated directly with the sewer. It should not be subject to corrosion or the possibility of sample contamination due to its materials of construction. The sample containers should be capable of being easily removed and cleaned; preferably they should be disposable."

Evaluation Parameters for Continuous Monitoring Equipment

"Evaluation of continuous monitors should address itself to: [2]
1. Vulnerability to obstruction or clogging of sampling ports, tubes and pumps.
2. Obstruction of main stream flow and susceptibility to damage.
3. Operation under the full range of flow conditions peculiar to storm and combined sewers.
4. Operation unimpeded by the movement of solids such as silt, sand, gravel and debris within the fluid flow; including durability.
5. Automatic start-up and operation (during storm conditions), unattended, self-cleaning.
6. Flexibility of operation allowed by control system.
7. Collection of samples of flotable materials, oils and grease, as well as coarser bottom solids.
8. Storage, maintenance and protection of collected samples from damage and deterioration as well as freedom of the sample train and containers from pre-contamination, adsorption, etc.
9. Amenability to installation and operation confined and moisture laden places such as sewer manholes.
10. Ability to withstand total immersion or flooding during adverse flow conditions.
11. Ability to withstand and operate under freezing ambiient conditions.
12. Ability to sample over a wide range of operating head conditions."

Many states have special monitoring requirements, such as those listed in the Appendix for the State of Florida. Additional requirements for monitoring have been imposed on sewage treatment plants by the Federal Government. Under Public Law 92-500, the National Pollutant Discharge Elimination System (NPDES) permit holders are required to submit a record of their monitoring results. [3]

    a. The permittee shall make and maintain records of all information resulting from the monitoring activities required by this permit.

    b. The permittee shall record for each measurement or sample taken pursuant to the requirements of this permit the following information: (1) the date, exact place, and time of sampling; (2) the dates analyses were performed; (3) who performed the analyses; (4) the analytical techniques or methods used; and, (5) the results of all required analyses.

c. If the permittee monitors any pollutant more frequently than is required by this permit, he shall include the results of such monitoring in the calculation and reporting of the values required in the Discharge Monitoring Report Form (EPA Form 3320-1 [10-72]). Such increased frequency shall be indicated on the Discharge Monitoring Report Form.

d. The permittee shall retain for a minimum of 3 years all records of monitoring activities and results including all records of calibration and maintenance of instrumentation and original strip chart recordings from continuous monitoring instrumentation. This period of retention shall be extended during the course of any unresolved litigation regarding the discharge of pollutants by the permittee or when requested by the Regional Administration or the State water pollution control agency.

6. Reporting of Monitoring Results

a. Monitoring information required by this permit shall be summarized and reported by submitting a Discharge Monitoring Report Form (EPA Form 3320-1 [10-72]), properly filled in and signed, to the Regional Administrator and the State agency at the following addresses:

U.S. Environ. Protection Agency　　　State Environ. Protection Agency
Region V, Permit Branch　　　　　　P. O. Box 1049
1 North Wacker Drive　　　　　　　Capital City, State
Chicago, Illinois 60606

b. Each submitted Discharge Monitoring Report shall be signed as follows:

(1) If submitted by a corporation, by a principal executive officer of at least the level of vice president, or his duly authorized representative, if such representative is responsible for the overall operation of the facility from which the discharge described in the Discharge Monitoring Report originates:

(2) If submitted by a partnership, be a general partner;

(3) If submitted by a sole proprietor, by the propietor;

(4) If submitted by a municipality, State or Federal agency, or other public entity, by a principal executive officer, ranking elected official, commanding officer, or other duly authorized employee.

c. All information submitted on the Discharge Monitoring Form shall be based upon measurements and sampling carried out during the three previous calendar months. The first Discharge Monitoring Report shall be submitted for a period ending 60 days from issuance. Thereafter, reporting periods shall end on the last day of each month. The permittee shall submit a Discharge Monitoring Report postmarked no later than the 28th day of the month following each completed reporting period.

BIBLIOGRAPHY

1. Federal Register, Vol. 39, No. 168, Wednesday, August 28, 1974, Water Quality & Pollutant Source Monitoring (Proposed Rules).
2. "An Assessment of Automatic Sewer Flow Samplers", Office of Research & Monitoring, USEPA, Washington DC 20460, EPA-R2-73-261, June 1973.
3. "Orientation to NPDES Permits", by Joseph F. Santner, National Training Center, Water Programs Operation, USEPA, Cincinnati, Ohio, 45268, EPA-430/1-74-013, December 1974.

APPENDIX

(A) Chapter 17—19 Rules of the State of Florida, Department of Pollution Control.
(B) NPDES Monitoring Report Form.
(C) Selected References on Monitoring.
(D) List of EPA Regional Offices.
(E) List of Manufacturers of Automatic Monitoring Equipment for Sewage Treatment Plants.

RULES OF THE STATE OF FLORIDA
DEPARTMENT OF POLLUTION CONTROL
CHAPTER 17-19
DOMESTIC WASTEWATER TREATMENT PLANT MONITORING

17-19.01 Purpose

17-19.02 Definitions

17-19.03 General Requirements — Violations

17-19.04 Sampling Methods

17-19.05 Sampling Schedules, Parameters

17-19.06 Sampling Points

17-19.07 Subsurface Disposal Wells

17-19.08 Disposal Ponds

17-19.09 Effective Date; Pre-existing Monitoring Contracts

17-19.01 Purpose. This rule chapter is adopted to assure compliance with the provisions of Chapter 17-4, Florida Administrative Code, with respect to wastewater treatment plants, and uniform quality control in sampling required of permit holders for such plants.

General Authority 403.061(6)(7)(15)(16) FS. Law Implemented 403.086, 403.078(2)(6), 403.101(2) FS. History-New 1-1-75.

17-19.02 Definitions. Definitions of technical terms used shall be in accordance with the latest edition of: Glossary — Water and Wastewater Control Engineering, American Public Health Association; Condensed Chemical Dictionary, Gessner G. Hawley, ed., Van Nostrand Rheinhold.

General Authority 403.061(6),(7),(15),(16) FS. Law Implemented 403.086, 403.078(2),(6), 403.101(2) FS. History-New 1-1-75.

17-19.03 General Requirements — Violations.

(1) The Permittee for any municipal or privately owned domestic wastewater

treatment plant, permitted under the provisions of Chapter 17-4, Florida Administrative Code, shall provide all analytical results detailed in this Chapter to the appropriate regional office of the Department of Pollution Control on a monthly basis. Such results shall be submitted in a timely manner so as to be received by the regional office by the fifteenth (15th) of the month following the month of sampling.

(2) Violation of any of the provisions of this Section shall be considered grounds for permit suspension or revocation under Chapter 17-4.10, Florida Administrative Code.

General Authority 403.061(6),(7),(15),(16) FS. Law Implemented 403.086, 403.078(2),(6), 403.101(2) FS. History-New 1-1-75.

17-19.04 Sampling Methods.

(1) Field testing, sample collection and preservation and laboratory testing, including quality control procedures, shall be in accordance with methods approved by the department and the United States Environmental Protection Agency. Copies of approved procedures shall be available for public inspection at the offices of the Department of Pollution Control. Copies shall be provided upon request for the cost of reproduction.

(2) As a quality control procedure, the department may require persons who submit test data to the department to analyze a reasonable number of reference samples which are to be provided by department or obtained by dividing samples with the operator.

(3) Where other testing procedures are found to be more satisfactory, such testing procedures will be used only upon prior approval of the department and the United States Environmental Protection Agency.

(4) The department may refuse to consider data obtained by unapproved procedures or methodology (including quality control procedures) as satisfying any requirements of this chapter.

General Authority 403.061(6),(7),(15),(16) FS. Law Implemented 403.086, 403.078(2),(6), 403.101(2) FS. History-New 1-1-75.

17-19.05 Sampling Schedules, Parameters.

(1) The minimum schedule for sampling and parameters to be sampled, for wastewater treatment plants without advanced waste treatment (as defined in Chapter 17-3.04(2) (b) 1 Florida Administrative Code), is as follows:

(2) The minimum schedule for sampling and parameters to be sampled, for wastewater treatment plants with advanced waste treatment (as defined in Chapter 17-3.04(2) (b) 1 Florida Administrative Code), is as follows:

(3) Schedules and parameters are subject to modification by the Department of Pollution Control's Executive Director or his designee as specified in construction or operation permits issued for wastewater treatment plants on a case by case basis, depending upon local requirements for assessment and maintenance of water quality.

General Authority 403.061(6),(7),(15),(16) FS. Law Implemented 403.086, 403.078(2),(6), 403.101(2) FS. History-New 1-1-75.

17-19.06 Sampling Points.

(1) Samples of both the influent and effluent are required.

(2) Plant influent shall be sampled between the grit chamber and the primary clarifier. The sample shall be collected so that it does not contain digester super-natant or returned activated sludge, or any other plant process recycled waters. Samples may usually be taken at a point between the grit chamber and the sludge return line provided that the above requirements are met.

(3) All effluent analyses shall be performed on samples collected immediately before discharge to surface or groundwaters except that the 5-day BOD and total suspended solids shall be determined based on samples taken before chlorination. If, however, chlorination is followed by sealed polishing ponds, samples for 5-day BOD and total suspended solids may be taken from the discharge from the pond.

(4) Sampling locations are subject to modification by the Executive Director or his designee as specified in construction or operation permits issued for wastewater treatment plants on a case by case basis, depending upon local requirements for assessment and maintenance of water quality.

General Authority 403.061(6),(7),(15),(16) FS. Law Implemented 403.086, 403.078(2),(6), 403.101(2) FS. History-New 1-1-75.

17-19.07 Subsurface Disposal Wells. In any case in which the effluent is discharg-ed to the subsurface through a well, sampling schedules and parameters identical to those used for surface receiving bodies of water are to be used for influent, effluent, and monitoring wells. Additional samples and analyses may be required by the Regional Administrator on a case by case basis as required to assess and maintain groundwater quality.

General Authority 403.061 (6),(7),(15),(16) FS. Law Implemented 403.086, 403.078(2),(6), 403.101(2) FS. History-New 1-1-75.

17-19.08 Disposal Ponds. In any case in which the effluent is released to a pond for disposal by percolation or evaporation, provision shall be made for at least semiannual sampling of the following parameters: 5-day BOD, chlorides, nitrate, nitrite, and fecal coliform. Such samples shall be taken from at least two shallow test wells located in the watertable hydraulically downstream from the pond. Ad-ditional sampling wells, samples and analyses may be required by the Regional Administrator on a case by case basis as required to assess and maintain ground-water quality.

General Authority 403.061 (6),(7),(15),(16) FS. Law Implemented 403.086, 403.078(2),(6), 403.101(2) FS. History-New 1-1-75.

17-19.09 Effective Date; Pre-existing Monitoring Contracts.

(1) This Chapter shall become effective on January 1, 1975.

(2) Notwithstanding any other provisions of this Chapter, domestic wastewater

Table 2. Minimum Sampling Schedules and Parameters for Wastewater Treatment Plants Without Advanced Waste Treatment.

Parameter	Design Capacity (mgd except where noted)					
	2,000 to 5,000 gpd (C)[1]	5,000 to 50,000 gpd (C)	0.05 to 0.5 (C)	0.5 to 1.0 (B or C)	1.0 to 5.0 (A,B,or C)	greater than 5.0 (A or B)
Flow (gpd or mgd)	daily	daily	daily	daily	continuous	continuous
pH[2]	none	none	daily	daily	daily	continuous
BOD (5 day)	quarterly	monthly	every two weeks	weekly	daily	daily
Suspended Solids	monthly	monthly	every two weeks	weekly	daily	daily
Fecal Coliform [2] (grab sample)	quarterly	quarterly	quarterly	monthly	every two weeks	weekly
Total N [2,5]	semiannually	semiannually	quarterly	monthly	every two weeks	weekly
Ammonia N [2,4,5]	none	none	quarterly	monthly	every two weeks	weekly
Nitrate N [2,5]	none	none	quarterly	monthly	every two weeks	weekly
Total Phosphate as P [2,4,5]	semiannually	semiannually	quarterly	monthly	every two weeks	weekly
Ortho Phosphate as P [2,4,5]	none	none	none	none	every two weeks	weekly
Total Chlorine Residual [2] (grab sample)	daily	daily	daily	daily	twice daily	continuous
Minimum Sampling Period [3]	grab	6 hours (grab up to 25,000 gpd [6])	8 hours (6 hrs. to 0.1 mgd)	12 hours (Class C - use 8 hours)	24 hours	24 hours

1 Facility classes of FAC Ch. 17-16 which includes the specified design capacity.
2 Required on effluent samples only.
3 To be composited over the specified period at hourly intervals.
4 Not necessary for land disposal system.
5 Parameters to be required for surface dischargers only when the Department determines that a potential water quality problem exists.
6 Sample to be taken at or near normal peak flow; sampling time to be reported.

Table 3. Minimum Sampling Schedules and Parameters for Wastewater Treatment Plants With Advanced Waste Treatment.

Parameter	Design Capacity (mgd except where noted)					
	2,000 to 5,000 gpd (C)[1]	5,000 to 50,000 gpd (C)	0.05 to 0.5 (C)	0.5 to 1.0 (B or C)	1.0 to 5.0 (A,B, or C)	greater than 5.0 (A or B)
Flow (gpd or mgd)	daily	daily	daily	daily	continuous	continuous
pH [2]	none	none	none	weekly	continuous	continuous
BOD (5 day)	monthly	every two weeks	weekly	two times per week	daily	daily
Suspended Solids	monthly	every two weeks	weekly	two times per week	daily	daily
Fecal Coliform [2] (grab sample)	quarterly	quarterly	quarterly	monthly	monthly	monthly
Total N [2]	every two weeks	every two weeks	weekly	two times per week	five times per week	five times per week
Ammonia N [2]	every two weeks	every two weeks	weekly	two times per week	five times per week	five times per week
Nitrate N [2]	every two weeks	every two weeks	weekly	two times per week	five times per week	five times per week
Total Phosphate as P when discharging to surface waters [2]	every two weeks	every two weeks	weekly	two times per week	five times per week	five times per week
Total Phosphate as P when [2] discharging to groundwaters	monthly	monthly	monthly	monthly	every two weeks	every two weeks
Free Chlorine Residual [2] (grab sample)	daily	daily	daily	daily	continuous	continuous
Minimum Sampling Period [3]	grab	6 hours (grab up to 25,000 gpd [4])	8 hours (6 hrs. to 0.1 mgd)	12 hours (Class C - use 8 hours)	24 hours	24 hours

1 Facility classes of FAC Ch. 17-16 which include the specified design capacity.
2 Required on effluent samples only.
3 To be composited over the specified period at hourly intervals.
4 Sample to be taken at or near normal peak flow; sampling time to be reported.

Your Treatment Plant Name

Your Address

City, State Zip Code

INSTRUCTIONS

1. Provide dates for period covered by this report in spaces marked "REPORTING PERIOD"
2. Enter reported minimum, average and maximum values under "QUANTITY" and "CONCENTRATION" in the units specified for each parameter as appropriate. Do not enter values in boxes containing asterisks. "AVERAGE" is average computed over actual time discharge is operating. "MAXIMUM" and "MINIMUM" are extreme values observed during the reporting period.
3. Specify the number of analyzed samples that exceed the maximum (and/or minimum as appropriate) permit conditions in the columns labeled "No. Ex." If none, enter "O".
4. Specify frequency of analysis for each parameter as No. analyses/No. days. (e.g., "3/7" is equivalent to 3 analyses performed every 7 days.) If continuous enter "CONT."
5. Specify sample type ("grab" or "___ hr. composite") as applicable. If frequency was continuous, enter "NA".
6. Appropriate signature is required on bottom of this form.
7. Remove carbon and retain copy for your records.
8. Fold along dotted lines, staple and mail Original to office specified in permit.

LL	00000000	001	N/A	N/A	N/A
ST	PERMIT NUMBER	DIS	SIC	LATITUDE	LONGITUDE

REPORTING PERIOD: FROM | 7 4 0 1 0 1 | TO | 7 4 0 2 0 1
YEAR MO DAY — YEAR MO DAY

PARAMETER		QUANTITY MINIMUM (3 card only)	AVERAGE	MAXIMUM	UNITS	NO. EX (4 card only)	CONCENTRATION MINIMUM	AVERAGE	MAXIMUM	UNITS	NO. EX	FREQUENCY OF ANALYSIS	SAMPLE TYPE
FLOW	REPORTED	.2	3.5	9	MGD		*****	*****	*****			1/1	GR
	PERMIT CONDITION	*****	4				*****	*****	*****			1/1	GR
PH	REPORTED	6	*****	7	STANDARD UNITS		*****	*****	*****			1/1	GR
	PERMIT CONDITION	6	*****	9			*****	*****	*****			1/1	GR
BOD 5 Infl/Effl	REPORTED		8555/400	10000/700	KG/DAY			600/30	800/45	MG/L		1/1	24
	PERMIT CONDITION	*****	550	825			*****	30	45			1/1	24
PERCENT REMOVAL BOD 5	REPORTED		95		%		*****	*****	*****			*****	*****
	PERMIT CONDITION		85	*****			*****	*****	*****			*****	*****
SUSPENDED SOLIDS Infl/Effl	REPORTED		8500/400	10000/700	KG/DAY			600/30	800/45	MG/L		1/1	12
	PERMIT CONDITION	*****	550	825			*****	30	45			1/1	12
PERCENT REMOVAL SUSPENDED SOLIDS	REPORTED		95		%		*****	*****	*****			*****	*****
	PERMIT CONDITION		85	*****			*****	*****	*****			*****	*****
FECAL COLIFORM	REPORTED	*****	*****	*****				150	315	N/100ML		1/1	GR
	PERMIT CONDITION	*****	*****	*****			*****	200	400			1/1	GRAB
	REPORTED												
	PERMIT CONDITION												

NAME OF PRINCIPAL EXECUTIVE OFFICER			TITLE OF THE OFFICER	DATE	I certify that I am familiar with the information contained in this report and that to the best of my knowledge and belief such information is true, complete, and accurate.	
Doe	John	B.	Mayor	7 4 0 2 1 3		SIGNATURE OF PRINCIPAL EXECUTIVE OFFICER OR AUTHORIZED AGENT
LAST	FIRST	MI	TITLE	YEAR MO DAY		PAGE OF

1/1 means that the sample was to be run daily. (Number of samples/number of days)
GR means grab sample.
Comp means composite sample and the number of hours over which the composite is made is the number which is entered

treatment plants, for which sampling required by the Department of Pollution Control is carried out by a contract in effect prior to December 16, 1974, may continue to sample in accordance with the requirement of any permit, certification, or rule of the Department until July 1, 1975 or the expiration of said contract, whichever occurs first.

General Authority 403.061(6),(7),(15),(16) FS. Law Implemented 403.086, 403.078(2),(6), 403.101(2) FS. History-New 1-1-75.

Flow is the rate at which waste water flows through a plant.

pH is a measure of acidity or alkalinity of a material, liquid, or solid. pH is represented on a scale of 0 to 14 with 7 representing a neutral state, 0 representing the most acid, and 14 the most alkaline.

Five Day Biochemical Oxygen Demand (BOD_5) is the amount of oxygen used by microorganisms in the biochemical oxidation of organic matter under specified conditions.

Suspended Solids (or Total Suspended Solids) are insoluable solids which were in suspension or dispersed in wastewater and were removed by some standard filtering procedure such as found on pages 278–279 in Methods for Chemical Analysis of Water and Wastes, 1971, EPA-NERC-AQCL, Cincinnati, Ohio 45268.

Fecal Coliforms are a group of bacteria common to the intestinal tracts of warm blooded animals. The presence of fecal coliform bacteria in water is an indicator of pollution.

Most Probable Number (MPN) is that number of organisms per 100 ml that, in accordance with statistical theory, would be more likely than any other number to yield the observed tests results.

$$\text{Percent removal} = \frac{\text{influent-effluent}}{\text{influent}} \times 100 \text{ assuming the flow is the same.}$$

If not the same, then weights must be used to obtain ratios; this seldom occurs.

Example: assume flow the same and influent $BOD_5 = 100$, effluent $BOD_5 = 15$

$$\text{Percent removal of } BOD_5 = \frac{100-15}{100} \times 100 = 85\%$$

SELECTED REFERENCES

1. Inter-Agency Committee on Water Resources Report No. 1, "Field Practice and Equipment Used in Sampling Suspended Sediment" (1940).
2. American Public Works Association — Research Foundation, "Water Pollution Aspects of Urban Runoff," Federal Water Pollution Control Administration, WPCR Series WP-20-15.
3. Ven Te Chow, *Open-Channel Hydraulics*, McGraw-Hill, New York (1959).
4. "Methods for Chemical Analysis of Water and Wastes — 1971" Environmental Protection Agency Report No. 16020 — 07/71.

5. Field, Richard and Struzeski, E. J., Jr., "Management and Control of Combined Sewer Overflows," *Journal Water Pollution Control Federation*, Vol. 44, No. 7 (1972).

6. American Society of Civil Engineers Task Committee on Sedimentation Research Needs Related to Water Quality, "Influences of Sedimentation on Water Quality: An Inventory of Research Needs," Vol. 97, No. HY8 (1971).

7. Inter-Agency Water Resources Council Report No. T, "Laboratory Investigation of Pumping Sampler Intakes" (1966).

8. "Strainer/Filter Treatment of Combined Sewer Overflows;" Fram Corporation; EPA Water Pollution Control Research Series Report No. WP-20-16:7/69.

9. "Stream Pollution Abatement from Combined Sewer Overflows;" Burgess and Niple, Ltd.; EPA Water Pollution Control Research Series Report No. DAST-32:11/69.

10. "Control of Pollution by Underwater Storage;" Underwater Storage, Inc.; EPA Water Pollution Control Research Series Report No. DAST-29:12/69.

11. "Engineering Investigation of Sewer Overflow Problems;" Hays, Seay, Mattern, and Mattern; EPA Water Pollution Control Research Series Report No. 11024 DMA 05/70.

12. "Microstraining and Disinfection of Combined Sewer Overflows;" Cochrane Division of the Crane Company; EPA Water Pollution Control Research Series Report No. 11023 EVO 06/70.

12a. "Microstraining and Disinfection of Combined Sewer Overflows — Phase II;" Environmental Systems Division of the Crane Company; EPA Environmental Protection Technology Series Report No. EPA-R2-73-124, January, 1973.

13. "Storm Water Pollution from Urban Land Activity;" AVCO Corporation; EPA Water Pollution Control Research Series Report No. 11034 FKL 07/70.

14. "Retention Basin Control of Combined Sewer Overflows;" Springfield Sanitary District; EPA Water Pollution Control Research Series Report No. 11023 — 08/70.

15. "Chemical Treatment of Combined Sewer Overflows;" The Dow Chemical Company; EPA Water Pollution Control Research Series Report No. 11023 FDB 09/70.

16. "Combined Sewer Temporary Underwater Storage Facility;" Melpar Company; EPA Water Pollution Control Research Series Report No. 11022 DPP 10/70.

17. "Urban Runoff Characteristics;" University of Cincinnati; EPA Water Pollution Control Research Series Report No. DQU 10/70.

18. "In-Sewer Fixed Screening of Combined Sewer Overflows;" Envirogenics Company, A Division of Aerojet General; EPA Water Pollution Control Research Series Report No. 11024 FKJ 10/70.

19. "Storm and Combined Sewer Pollution Sources and Abatement;" Black, Crow, and Eidsness, Inc.; EPA Water Pollution Control Research Series Report No. 11024 ELB 01/71.

20. "Storm Water Problems and Control in Sanitary Sewers;" Metcalf and Eddy, Inc.; EPA Water Pollution Control Research Series Report No. 11024 EQG 03/71.

21. "Underwater Storage of Combined Sewer Overflows;" Karl R. Rohrer Associates, Inc.; EPA Water Pollution Control Research Series Report No. 11022 ECV 09/71.

22. "Maximizing Storage in Combined Sewer Systems;" Municipality of Metropolitan Seattle; EPA Water Pollution Control Research Series Report No. 11022 ELK 12/71.

23. Inter-Agency Committee on Water Resources Report No. 5, "Laboratory Investigation of Suspended Sediment Samplers" (1941).

24. Inter-Agency Water Resources Council Report No. Q, "Investigation of a Pumping Sampler With Alternate Suspended Sediment Handling Systems" (1962).

25. E. F. Schulz, R. H. Wilde, and M. L. Albertson, "Influence of Shape on the Fall Velocity of Sedimentation Particles," Missouri River Division Sedimentation Series Report No. 5, U.S. Army Corps of Engineers, Omaha, Nebraska (1954).

26. Inter-Agency Committee on Water Resources Report No. 12, "Some Fundamentals of Particle Size Analysis" (1957).

27. A. M. Mkhitaryan, *Gidravlika i Osnovy Gazodinamiki,* Gosudarstvennoe Izdatel 'stvo Tekhnicheskoi Literatury UkrSRR, Kiev (1959).

28. V. S. Knorz, "Beznapornyi gidrotransport i ego raschet," Izvestiya Vsesoyuznogo Naucho-Issledovatel 'skogo Instituta Gidravliki, Vol. 44 (1951).

29. Water Pollution Control Federation Manual of Practice No. 9, *Design and Construction of Sanitary and Storm Sewers* (1970).

REGIONAL MUNICIPAL PERMIT COORDINATORS

Mr. Frank Ciaveterri, Chief
Municipal Permit Section
Municipal Facilities Branch
Air and Water Programs Division
EPA — Region I
JFK Federal Building
Boston, Massachusetts 02203
8 (617) 223-5077

Mr. David Luoma, Director
Water Division
EPA — Region II
26 Federal Plaza (Foley Square)
New York, New York 10007
8 (212) 264-2513

Mr. Joseph Galda, Chief
Municipal Facilities
Enforcement Division
EPA — Region III
Sixth and Walnut Streets
Philadelphia, Pennsylvania 19106
8 (215) 597-0402

Mr. Roger D. Graham
Water Programs Office
EPA — Region IV
1421 Peachtree Street, N.E.
Atlanta, Georgia 30309
8 (404) 528-3012

Mr. Clarence Laskowski, Chief
Operation and Maintenance Section
Construction Grants Branch
Water Division
EPA — Region V
1 North Wacker Drive
Chicago, Illinois 60606
8 (312) 353-1052

Dr. Richard L. Hill, Director
Air and Water Programs Division
EPA — Region VI
1600 Patterson Street, Suite 1100
Dallas, Texas 75201
8 (214) 749-1267

Dr. Dennis A. Degner, Chief
Municipal Permits Section
Enforcement Division
EPA — Region VII
1735 Baltimore Avenue
Kansas City, Missouri 64108
8 (816) 374-5955

Mr. Charles Murray, Jr., Director
Water Division
EPA — Region VIII
1860 Lincoln Street
Denver, Colorado 80203
8 (303) 837-3961

Mr. Robert Rock
Municipal Permits Section
Hazardous Materials Control Division
EPA — Region IX
100 California Street
San Francisco, California 94111
8 (415) 555-3450

Mr. Robert Stamnes, Chief
Municipal Permit Section
Enforcement Division
EPA — Region X
1200 Sixth Avenue
Seattle, Washington 98101
8 (205) 442-1280

REGIONAL PERMIT PROGRAM CHIEFS

Mr. E. J. Conley
Chief, Permits Branch
EPA — Region I
JFK Federal Building
Boston, Massachusetts 02203
(617) 223-5061

Dr. Ernest Regna
Chief, Permits Branch
EPA — Region II
Room 3914F
26 Federal Plaza
New York, New York 10007
(212) 264-9894

Mr. William J. Riley
Chief, Permits Branch
EPA — Region III
Curtis Building
6th and Walnut Streets
Philadelphia, Pennsylvania 19106
(215) 597-9380

Mr. George Harlow
Chief, Water Enforcement Branch
EPA — Region IV
1421 Peachtree Street, N.E.
Atlanta, Georgia 30309
(404) 526-5201

Mr. Al Manzardo
Chief, Permits Branch
EPA — Region V
1 North Wacker Drive
Chicago, Illinois 60606
(212) 353-1476 or 1232

Dr. Ned Burleson
Chief, Permits Branch
EPA — Region VI
1600 Patterson Street, Suite 1100
Dallas, Texas 75202
(214) 749-1983

Mr. Carl Walter
Chief, Permits Branch
EPA — Region VII
1735 Baltimore Avenue
Kansas City, Missouri 64108
(816) 374-5955

Mr. Evan Dildine
Chief, Permits Branch
EPA — Region VIII
Suite 900, Lincoln Tower
1860 Lincoln Street
Denver, Colorado 80203
(303) 837-4901

Mr. Bill Pierce
Chief, Permits Branch
EPA — Region IX
100 California Street
San Francisco, California 94111
(415) 556-3450

Mr. Lloyd Reed
Chief, Permits Branch
EPA — Region X
1200 6th Avenue
Seattle, Washington 98101
(206) 442-1213

WHERE TO GET FURTHER INFORMATION

In order to get details or any other aspects of the Technology Transfer Program, contact your EPA Regional Technology Transfer Committee Chairman from the list below:

Region	Chairman	Address
I	Lester Sutton	Environmental Protection Agency John F. Kennedy Federal Building Room 2304 Boston, Massachusetts 02203 617 223-2226 (Maine, N.H., Vt., Mass., R.I., Conn.)

(continued)

II	Rocco Ricci	Environmental Protection Agency 26 Federal Plaza New York, New York 10017 212 264-2513 (N.Y., N.J., P.R., V.I.)
III	Kenneth Suter	Environmental Protection Agency 6th and Walnut Streets Philadelphia, Pennsylvania 19106 215 597-9268 (Pa., W.Va., Md., Del., D.C., Va.)
IV	Asa B. Foster, Jr.	Environmental Protection Agency Suite 300 1421 Peachtree Street, N.E. Atlanta, Georgia 30309 404 526-3454 (N.C., S.C., Ky., Tenn., Ga., Ala., Miss., Fla.)
V	Clifford Risley	Environmental Protection Agency 1 N. Wacker Drive Chicago, Illinois 60606 312 353-5756 (Mich., Wis., Minn., Ill., Ind., Ohio)
VI	Jocelyn G. Kempe	Environmental Protection Agency 1600 Patterson Street, Suite 1100 Dallas, Texas 75201 214 749-1238 (Texas, Okla., Ark., La., N. Mex.)
VII	Lewis A. Young	Environmental Protection Agency 1735 Baltimore Avenue Kansas City, Missouri 64108 816 374-2725 (Kansas, Nebr., Iowa, Mo.)
VIII	Russell Fitch	Environmental Protection Agency 1860 Lincoln Street Denver, Colorado 80203 303 837-3849-837-3691 (Colo., Mont., Wyo., Utah, N.D., S.D.)
IX	Frank Covington	Environmental Protection Agency 100 California Street San Francisco, California 94111 415 556-0218 (Calif., Ariz., Nev., Hawaii)
X	John Osborn	Environmental Protection Agency 1200 6th Avenue Seattle, Washington 98101 206 442-1296 (Wash., Ore., Idaho, Alaska)

U.S. ENVIRONMENTAL PROTECTION AGENCY
REGIONAL MANPOWER OFFICERS

REGION I

Mr. Edgar Bernard, Chief (Lee Siegel, Secy)
Manpower Development Branch
John F. Kennedy Federal Building
Boston, Massachusetts 02203
    (617) 223-5765

REGION II

Mr. Robert Knox, Chief (Iris Rivera, Secy)
Manpower Development and Trng. Office
Air and Water Programs
26 Federal Plaza
New York, New York 10007
    (C. Tenerella) (212) 264-1316

REGION III

Mr. James Kennedy, Act. Chief (Neena, Secy)
Manpower Development Office
Air and Water Programs
Curtis Building
6th and Walnut Strrets
Philadelphia, Pennsylvania 19106
    (215) 597-9297

REGION IV

Mr. Bob Laughran, Acting Chief,
  404-526-5975
Manpower Development Branch
Division of Air and Water Programs
1421 Peachtree Street, NE, Fourth Floor
Atlanta, Georgia 30309
    R. Towner - SEWL (404) 546-3161

REGION V

Mr. C. J. Shura, Chief
Manpower Development Branch
Office of Air and Water Programs
1 N. Wacker Drive
Chicago, Illinois 60606
    (Mullineaux) (312) 353-1456 or 1457

REGION VI

Ms. Joceyln Kempe, Acting Chief
Manpower Development Branch
Air and Water Programs Division
1600 Patterson
Dallas, Texas 75201
    (214) 749-1885

REGION VII

Mr. Jack Coakley, Chief
Manpower Development Branch
Air and Water Programs
1735 Baltimore
Kansas City, Missouri 64108
    (Steve Fishman) (816) 374-5971

REGION VIII

Mr. Elmer Chenault, Chief
Manpower Development Branch
Air and Water Division
1860 Lincoln Street - 9th Floor
Denver, Colorado 80203
    (Charles Stevens) (303) 837-4343

REGION IX

Dr. William Biship, Chief
Manpower Development Branch
Air and Water Division
100 California Street
San Francisco, California 94111
(Joe Canata)   (415) 556-4806
    Mr. Melvin Koizumi
    Honolulu Office, EPA, Region IX
    The Bishop Trust Building
    1000 Bishop Street
    Honolulu, Hawaii 96813
    (808) 546-8910

REGION X

Mr. Robert Courson, Chief
Manpower and Training Branch
Division of Air and Water Programs
1200 Sixth Avenue - Mail Stop 345
Seattle, Washington 98101
    (206) 442-1296

EPA REGIONAL PUBLIC AFFAIRS

*The Public Affairs Director in each Regional Office can provide assistance and materials to individuals and groups who are seeking to work on water pollution control problems.

<u>REGIONAL OFFICES</u>

<u>STATES COVERED</u>

Mr. Paul Keough
EPA Region I
John F. Kennedy Building
Boston, Massachusetts 02203
(617) 223-7223

Connecticut, Maine, Massachusetts, New Hampshire, Rhode Island, Vermont

*Mr. Donald Bliss
EPA Region II
26 Federal Plaza
New York, New York 10007
(212) 264-2515 or *9895

New Jersey, New York, Puerto Rico, Virgin Islands

Mr. Gary Brooton
EPA Region III
6th and Walnuts
Philadelphia, Pennsylvania 19106
(215) 597-9904 or *9370

Delaware, Maryland, Pennsylvania, Virginia, West Virginia, D.C.

Mr. Charles Pou
EPA Region IV
1421 Peachtree Street, N.E.
Atlanta, Georgia 30309
(404) 526-3004

Alabama, Florida, Georgia, Kentucky, Mississippi, North Carolina, South Carolina, Tennessee

*Mr. Frank Corrado
EPA Region V
One North Wacker Drive
Chicago, Illinois 60606
(312) 353-1478 or *5800

Illinois, Indiana, Michigan, Minnesota, Ohio, Wisconsin

*Ms. Betty Williamson
EPA Region VI
1600 Patterson Street
Dallas, Texas 75201
(214) 749-1151 or *1962

Arkansas, Louisiana, New Mexico, Oklahoma, Texas

Mr. Randall Jessee
EPA Region VII
1735 Baltimore Avenue

Iowa, Kansas, Missouri, Nebraska

(continued)

Kansas City, Missouri 64108
(816) 374-5894

*Mr. Howard Kayner
EPA Region VIII
1860 Lincoln Street
Denver, Colorado 80203
(303) 837-4905 or *4904

Colorado, Montana, North Dakota,
South Dakota, Utah, Wyoming

Mr. Allan Abramsor
Chief, External Intergovernmental
  Affairs Branch
EPA Region IX
100 California Street
San Francisco, California 94111
(415) 556-6266

Arizona, California, Hawaii, Nevada,
American Samoa, Guam, Trust
Territories of the Pacific, Wake
Island

Mr. Robert Jacobson
EPA Region X
1200 Sixth Avenue
Seattle, Washington 98101
(206) 442-1203

Alaska, Idaho, Oregon, Washington

APPENDIX E

BIF Sanitrol
P. O. Box 41
Largo FL 33540

Brailsford & Co., Inc.
Milton Road
Rye, NY 10580

Brandywine Valley Sales Co.
P. O. Box 243
Honeybrook PA 19344

Chicago Pump Division
FMC Corporation
622 Diversey Parkway
Chicago IL 60614

Hydra-Numatic Sales Co.
65 Hudson St.
Hackensack NJ 07602

Westinghouse Electric Corp.
Infilco Division
Box 2118
Richmond VA 23216

Instrumentation Specialties Co., Inc.
P. O. Box 5347
Lincoln, Neb 68524

Lakeside Equipment Corp.
1022 East Devon Avenue
Bartlett, IL 60103

Markland Specialty Engineering Ltd.
Box 145
Etobicoke, Ontario, Canada

N-Con Systems Co., Inc.
308 Main St.
New Rochelle NY 10801

(continued)

Phipps and Bird, Inc.
303 South 6th Street
Richmond, VA 23205

Protech, Inc.
Roberts Lane
Malvern, PA 19355

Quality Control Equipment Co.
6139 Fleur Drive
Des Moines, IO 50315

Sonford Products Corp.
100 East Broadway, Box B
St. Paul, MI 55071

Sigmamotor, Inc.
14 Elizabeth St.
Middleport, NY 14105

Sirco Controls Co.
8815 Selkirk St.
Vancouver, B.C., Canada

Testing Machines, Inc.
400 Bayview Avenue
Amityville, NY 11701

CHAPTER 5

ENERGY OPERATIONS PLANS

INTRODUCTION

An episode at a sewage treatment plant (STP) is when through an Act of God, a malfunction or an equipment failure, untreated or partially treated waste reaches a wastercourse in excess of:

- Water Quality Limitation
- Waste Load Allocations
- Design Capacity of the STP

The entire aim of sewage treatment is to: [1]

"1. Maintenance of sources for use as domestic water supplies

2. Prevention of disease

3. Prevention of nuisances

4. Maintenance of clean waters for bathing and other recreational purposes

5. Maintenance of clean waters for the propagation and survival of fish life

6. Conservation of water for industrial and agricultural uses

7. Prevention of silting in navigable channels"

Under normal operation, a STP is designed to meet these objectives. When some unforeseen occurence takes place in the system, the operator should be flexible enough to handle this problem. Unfortunately, very little has been written concerning episodes. The general philosophy during an upset has been to:

1. Eliminate or correct the problem as quickly as possible

2. To return the plant to normal operation as soon as possible

In many instances, the operator is faced with discharge/bypass and/or repair. This, then, becomes a trade-off between the time to correct the problem versus the down time before the sewers start to surcharge (back up into homes and commercial establishments).

Reporting of Episodes

Both State and Federal agencies are requesting that adverse environmental impact, by-passing and power failures be reported. A typical Federal NPDES permit addresses episodes as follows: [2]

"Noncompliance Notification

"If, for any reason, the permittee does not comply with or will be unable to comply with any daily maximum effluent limitation specified in this permit, the permittee shall provide the Regional Administrator and the State with the following information, in writing, within five (5) days of becoming aware of such condition:

"a. A description of the discharge and cause of noncompliance; and

"b. The period of noncompliance, including exact dates and times; or, if not corrected the anticipated time the noncompliance is expected to continue, and steps being taken to reduce, eliminate and prevent recurrence of the noncomplying discharge.

"Facilities Operation

"The permittee shall at all times maintain in good working order and operate as efficiently as possible all treatment or control facilities or systems installed or used by the permittee to achieve compliance with the terms and conditions of this permit.

"Adverse Impact

"The permittee shall take all reasonable steps to minimize any adverse impact to navigable waters resulting from noncompliance with any effluent limitations specified in this permit, including such accelerated or additional monitoring as necessary to determine the nature and impact of the noncomplying discharge.

"Bypassing

"Any diversion from or bypass of facilities necessary to maintain compliance with the terms and conditions of this permit is prohibited, except (i) where unavoidable to prevent loss of life or severe property damage, or (ii) where excessive storm drainage or runoff would damage any facilities necessary for compliance with the effluent limitations and prohibitions of this permit. The permittee shall notify the permit issuing authority in writing within 72 hours of each such diversion or bypass in accordance with the procedures specified above for reporting noncompliance. The permittee shall within 30 days after such incident submit to EPA for approval a plan to prevent reccurrence of such incidents.

"Power Failures

"The permittee is responsible for maintaining adequate safeguards to prevent the discharge of untreated or inadequately treated wastes during electrical power failures either by means of alternate power sources, stand-by generators or retention of inadequately treated effluent. Should the treatment works not include the above capabilities at time of permit issuance, the permittee must furnish within 120 days to the permitting authority, for approval, an implementation schedule for their installing."

State permit requirements are very similar to the Federal and require the permittee to report and to describe the incident and what corrective measures were taken. Typical of this requirement is an excerpt from the State of Florida Regulations: [3]

17-4.13 Plant Operation — Problems.

In the event the permittee is temporarily unable to comply with any of the conditions of the permit due to breakdown of equipment or destruction by hazard of fire, wind or by other cause, the permittee is to immediately notify this Department. Notification shall include pertinent information as to the cause, and what steps are being taken to correct the problem and prevent its recurrence and the owner's intent toward reconstruction of destroyed facilities where applicable.

General Authroity 403.021, 403.031, 403.061, 403.088 FS. Law Implemented 403.021, 403.031, 403.061, 403.087, 403.088 FS. History—New 3-4-72, Revised 5-17-72.

CAUSES OF EMERGENCIES [4]

Emergencies at a wastewater treatment pland do not just happen — they are caused. The individual causes are sometimes numerous; however, many emergencies have certain similarities, particularly when they are traced back to their origin. For purposes of this manual, all causes of emergencies have been grouped under, or related to, one of the following:

"Natural Disasters

An event, concentrated in time and space, which causes a community or a facility to suffer such damage as to disrupt its normal functions and operations can be termed a natural disaster.

"Natural disasters which are most likely to affect the operation of a wastewater treatment facility to the extent of reducing the efficiency of the plant can be associated with one of the following:

      Hurricane
      Tsunami (tidal wave)
      Blizzard
      Forest and grass fire
      Tornado
      River flood
      Earthquake

"A study should be made to determine the potential for natural disasters in the areas where the municipal wastewater treatment system is located. Information on the natural disasters previously mentioned is available from agencies such as the U.S. Army Corps of Engineers and the Office of Civil Defense, Departments of Interior and Agriculture, and the Departments of Commerce and Transportation. Many state and local agencies, as well as volunteer disaster relief organizations such as the American National Red Cross, also have compiled information on disasters.

"Civil Disorders and Strikes

People have been demonstrating against war, for and against various social movements such as housing, civil rights, women's liberation, etc., for a very long time. In more recent times, workers have banded together to form unions and have subsequently used the strike as an effective tool in making their demands heard.

"According to the U.S. Treasury Department, there were over 5,800 bombings or attempted bombings during the 15-month period ending April 1970. As a result of these acts, 42 persons lost their lives, 384 were injured, and over $20 million of property damage was reported.

"The interrelationship of these types of events — civil disorders and strikes — with the operation of wastewater treatment facilities poses a new type of threat. The importance of uninterrupted treatment of wastewater is of primary concern because of the ever-increasing need for pollution-free waters.

"Several facets of widespread civil disorder might well be the destruction of a sewage pumping station, the bombing of a power substation, or the dumping of toxic material into a manhole. Any one of these actions could interrupt the normal operation of a wastewater treatment facility and subsequently lead to pollution of nearby waters.

"Faulty Maintenance

Equipment manufactured by man to help him in his everyday tasks is generally similar in one respect; it must be maintained or the equipment will cease to perform the task for which it was constructed. The manner in which equipment is maintained will generally determine how well it will perform its intended function and for how long. Good maintenance will result in equipment performing throughout its design period; however, poor maintenance or faulty maintenance will shorten the expected life of equipment.

"Unexpected breakdowns due to faulty maintenance can greatly affect the continued operation of a wastewater treatment plant. Although the breakdown can possibly be repaired during a regularly scheduled repair program and probably does not represent an emergency, it is the effect on the continued satisfactory operation of the plant that can lead to the emergency condition.

"Negligent Operation

All operations, regardless of application, large or small, require that certain procedures be followed for satisfactory performance. Applying this rule to a wastewater treatment plant's operation is no different. The operations required for the plant to function in a satisfactory manner require that certain procedures be followed, whether the procedures were established in-house, by a regulatory agency, or by the local governing body. To improperly follow established procedures constitutes negligent operation.

"In many instances, negligent operation may not be as readily noticeable as faulty maintenance, but the emergency condition resulting from it could possibly be more severe because negligent operation could affect more units of operation before being discovered.

"It is therefore imperative that sound operating procedures be developed and maintained to ensure the satisfactory operation of all wastewater treatment plants.

"Accidents

The old saying "accidents don't just happen — they are caused" certainly applies when we speak of accidents relative to wastewater treatment plants and emergency conditions. Generally speaking, accidents result in personal injury and property damage, both of which have a direct bearing on a plant's operation. However, for this manual's use, an additional effect of accident, concern is directed to the emergency condition arising from the accidental spill of toxic substance into the sewerage system. This one accident, if it goes undetected long enough, could shut down process units of the largest treatment plants for a considerable length of time, thus causing a severe emergency condition."

Table 1.

CAUSES	EFFECTS	REASONS FOR EFFECTS
Natural Disaster	Personnel Absence	Flooding
Civil Disorder		Wind
		Fire
		Injury
		Picket Lines
		Blocked Access
Natural Disaster		Flooding
Civil Disorder		Structural Damage
Faulty Maintenance	Equipment Failure	Loss of Power
Negligent Operation		Sabotage
Accidents		Overload Condition
		Clogged Pipelines
		Overheating
Natural Disaster		Flooding
Civil Disorder		Wind
Faulty Maintenance	Power Loss	Sabotage
Negligent Operation		Salt Spray on Lines
Accidents		Structural Damage
		Ice on Lines

Table 1. (continued).

		Fire
Natural Disaster	Blocked Access	Flooding
Civil Disorder		Wind
		Slides
		Debris
		Road Washouts
		Fire
		Sabotage
Natural Disaster	Communications Loss	Flooding
Civil Disorder		Wind
		Sabotage
		Fire
Natural Disaster		Flooding
Civil Disorder		Loss of Power
Faulty Maintenance	Process Failure	Sabotage
Negligent Operation		Personnel Absence
Accidents		Toxic Spill

STP Design Considerations

A lot of emphasis should be placed on the design of the STP to provide *flexibility* in operation and *standby* equipment to cope with episodes.

The major components in the design of a facility for episode control normally considered are:

- By-pass lines
- Chlorinators to super-chlorinate if by-passing is required
- Auxiliary power
- Piping flexibility to enable circumvention of certain operations during episodes.

The familiar 10 states standards includes provisions for: [5]

"35. Emergency Operation

"35.1 Objective

The objective of emergency operation is to prevent the discharge of raw or partially treated sewage to any waters and to protect public health by preventing back-up of sewage and subsequent discharge to basements, streets, and other public and private property.

"35.2 Emergency Power Supply

Provision of an emergency power supply for pumping stations should be made, and may be accomplished by connection of the station to at least 2 independent public utility sources, or by provision of portable or in-place internal combustion engine equipment which will generate electrical or mechanical energy, or by the provision of portable pumping equipment.

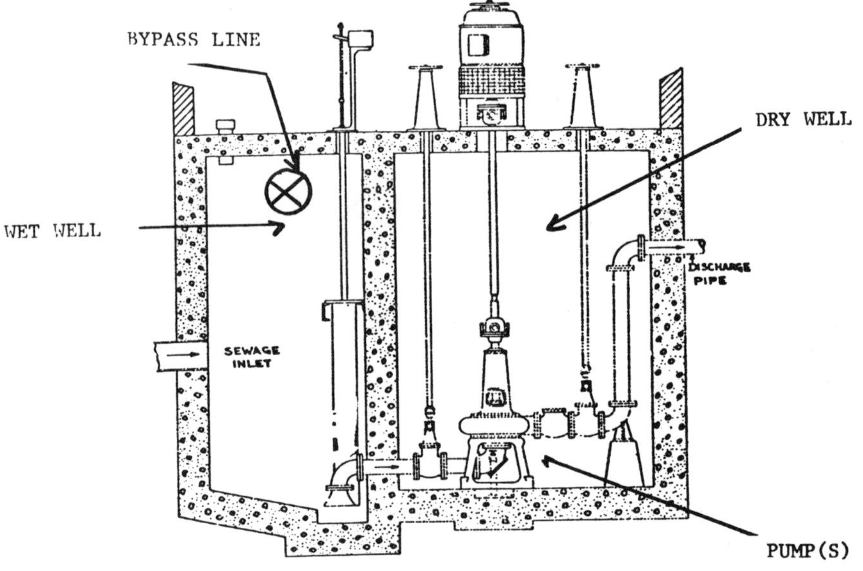

Figure 1. A dry-Pit Lift Station.

"35.21 In-place Equipment

Where in-place internal combustion equipment is utilized, the following guidelines are suggested for use:

35.211 Placement

The unit shall be bolted in place. Facilities shall be provided for unit removal for purposes of major repair or routine maintenance.

35.212 Controls

Provision shall be made for automatic and manual start-up and cut-in.

35.213 Size

Unit size shall be adequate to provide power for lighting and ventilation systems and such further systems affecting capability and safety.

35.214 Engine Location

The unit internal combustion engine shall be located above grade with suitable and adequate ventilation of exhaust gases.

"35.22 Portable Equipment

Where portable equipment is utilized the following guidelines are suggested for use:

Pumping units shall have capability to operate between the wet well and the discharge side of the station, with the station provided with permanent fixtures which will facilitate rapid and easy connection of lines.

Electrical energy generating units shall be protected against burn-out when normal utility services are restored, and shall have sufficient capacity to provide power for lighting and ventilation systems and such further station systems affecting capability and safety.

"35.23 Emergency Power Generation

All emergency power generation equipment should be provided with instructions indicating the essentiality of routinely and regularly starting and running such units at fill load.

"35.3 Overflows

The provision of a high-level wet well overflow to supplement alarm systems and emergency power generation should be considered. Where a high level overflow to supplement alarm systems and emergency power generation should be considered. Where a high level overflow is utilized, consideration shall also be given to the installation of storage-detention tanks, or basins, which shall be made to drain to the station wet well. Where such overflows affect public water supplies, shell-fish production, or waters used for culinary or food processing purposes, a storage-detention basin, or tank, shall be provided having 2 hour detention capacity at the anticipated overflow rate.

"72.3 Settling Tanks

The following requirements are in addition to those set forth in Section 54:

72.31 By-Pass

When a primary settling tank is used, provision shall also be made for discharging raw sewage directly to the aeration tanks following pretreatment."

The following chart indicates some of the types of failures in sewage systems and some of the design considerations that should be incorporated into the plant.

SEWAGE TREATMENT PLANT OPERATING CONSIDERATIONS

The following is an excerpt from the Federal Guidelines [6] for the operation of a sewage treatment plant.

"*6.0 Emergency Operating Plan.*

"6.1 To protect the health and welfare of municipal wastewater treatment plant personnel, and to minimize adverse effects in times of emergencies, wastewater treatment facilities constructed under P.L 92-500 grants should have included in the operation and maintenance manual, a section establishing a comprehensive plan for emergency operating procedures.

Table 2. Emergency Episode Considerations.

	TYPE OF FACILITY/OPERATION	TYPE OF EMERGENCY	DESIGN CONSIDERATIONS FOR PREVENTION	OPERATION CONSIDERATIONS FOR CORRECTION
COLLECTION SYSTEM	Lift Station	Electrical Failure	Stand by Generator	Start generator & call electrical repair service
		Pump Failure	Dual Pumps	Inspection & spare parts inventory
			By-Pass Line	Chlorinate at lift station during by-passing
	Force Main	Lift Rupture	By-Pass at Lift Station	Chlorinate at lift station during by-passing
				Call scavenger to pump out wet well
				Repair as quickly as possible
SEWAGE TREATMENT PLANT	Comminutor	Mechanical or Electrical Failure	Bar Screen Prior to Comminutor	Insert temporary screen to protect equipment downstream
	Excessive Storm Water	Hydraulic overload or flooding	By-Pass Line	By-Pass until surge subsides
			Separate Storm Sewers	-----
	Electrical Equipment	Power Failure	By-Pass Line	Start Stand-by power & call power company
			Stand-By Generator	
			Dual Power Lines	
	Biological Treatment Units	Toxic Materials or Industrial Waste	Early Warning System	Divert flow around sensitive biological treatment unit or to holding pond for batch treatment
			By-Pass System to Ponds	
			By-Pass Line	
	Digester	Digester Malfunction (Digester goes sour or inert due to operation or toxic materials)	Dual Digesters	Isolate digester & chemically treat
			Internal Recirculation System	

"6.2 Wastewater Treatment equipment suppliers should include emergency operating instructions with all equipment. This will enable the consulting engineer to incorporate this information as he prepares the operation and maintenance manual. Also, the consulting engineers, using emergency equipment instructions, may make an evaluation of equipment with regard to flexibility during emergencies. An evaluation of this type will allow plant personnel to respond more efficiently to emergencies affecting the equipment.

"6.3 The plan should insure the most effective operation possible under emergency conditions.

"6.4 The plan should protect the waste treatment facilities under all foreseeable emergency conditions. It should be complete and comprehensive and should include, but not be limited to, the following:

    a.   Effects of Emergencies

    b.   Vulnerability Analysis of the System

    c.   Protective Measures

    d.   Responses to Emergencies

    e.   Emergency Response Program

"6.5 The emergency operating plan must be periodically updated to insure current measures and responses are valid. Mutual aid agreements and notification procedures may change and must be validated periodically to enable the emergency operating plan to function properly."

This first assumes that an emergency operating plan will be available. On new, Federally funded facilities this is a requirement under P.L. 92-500, but does not apply to those plants already in operation. Any STP should have a plan to follow in case of emergencies, and have his readily available to the operators. One of the first steps in deciding what failures may occur is a vulnerability analysis. [6]

"A municipal wastewater treatment system includes the collection lines, pump stations, and treatment plant. The system requires trained personnel, power, materials, supplies, and communications to function properly. It is essential to estimate both the strengths and weaknesses of an individual system in relation to anticipated emergency conditions prior to drafting an emergency operating plan. A vulnerability analysis of the system is an estimation of the degree to which the system is adversely affected, in relation to the function it must perform, by an emergency condition. This analysis should include power supply, communications, equipment, material, supplies, personnel, security, and emergency procedures. After performing a vulnerability analysis upon a given system for several possible emergency conditions and then comparing the results of the various analyses, certain key — most vulnerable — components of the system can be identified. The following steps should be followed in making a vulnerability analysis:

1. List components of treatment system.

2. Select emergency condition to be investigated.

3. Estimate effects of emergency condition on each component of system; use

vulnerability worksheet.

4. Estimate treatment system's ability to perform its intended function during the emergency.

5. Identify key system components responsible for the failure when a system fails to perform.

"When identifying those components which are partially or totally incapacitated by the emergency condition, attention should be given to those system components which are interrelated with other components so as to make the entire system inoperative. These components are the most vulnerable. The emergency operating plan should indicate priorities of repair of the system and alternate provisions in case of light or severe damage. The following methods can be employed to reduce the system's vulnerability:

1. An optimum preventive maintenance and testing program.
2. Duplication and separation of vital works
3. Minimizing dependence on power and pumping
4. Provision for more than one power source and/or transmission line
5. Flexibility in operation of treatment works
6. Maintenance of adequate chemical supplies
7. Provision of dual power sources, on-site storage of fuel and auxiliary power units, remote and/or automated controls, and ready conversion of automatic controls to manual operation
8. Provision of portable pumps with fuel-operated units
9. Provision at major pumping stations of more than one incoming and discharge line
10. Training of regular and auxiliary personnel in emergency operations and procedures
11. Conducting emergency operations exercises periodically"

Once a vulnerability analysis is completed, a warning system must be installed to indicate failure and steps must be prescribed by which to take corrective measures.

"Warning Devices [6]

All wastewater treatment systems regardless of size need warning devices of some type. The small systems require less sophisticated equipment than the large systems; however, the need is always present to warn the man on duty of impending or existing emergencies.

"There are many individual types of devices available but generally they can be classified as follows:

1. Alarms — which may be audio-visual
2. Indicating lights — tell the operator which equipment is required to run, equipment running, power on and off, etc.
3. Indicators — mechanical (gauges), electical (counters, indicators, recorders) or electro-mechanical (flow and/or pressure recorders)

"Many distinct uses can be made of warning devices; the following are just a few:
  1. Low pressure (discharge, suction, or level)
  2. No pressure (discharge, suction, or level)
  3. Low well water level
  4. Flooding
  5. Air in pump
  6. Air loss
  7. Power failure (transfer to standby source)
  8. Equipment failure
  9. Pump reversal
 10. Freezing

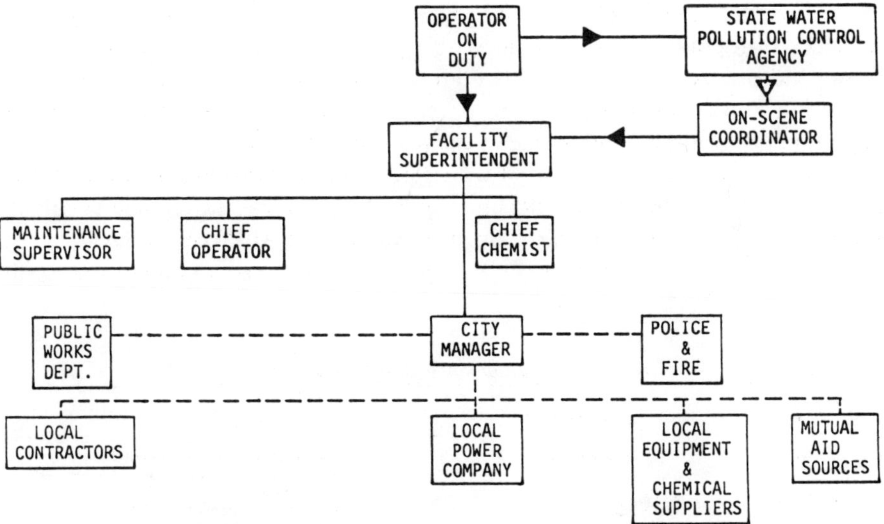

Figure 2. Emergency Condition Flow Diagram [6].
(Designation & Notification System)

Notification System

Somewhere between the warning of an episode occurance and the correction of the problem is the alert program for the notification of personnel; those people involved with the emergency (utility people, contractors, etc.) who will need to be on site during this period.

These specific individuals will then be responsible to handle the emergency.

Corrective Measures

Each person involved in an emergency plan should be aware of:

- Equipment and inventory available
- How to operate equipment or treatment units during various episode conditions.
- Availability of additional assistance.

Pretreatment of Wastes

One method of preventing discharges into a STP that will cause an emergency to a biological system is to restrict the allowable discharges into the sewer system. For many years, municipal ordinances have contained provisions restricting certain materials in sewers. Typical of this type of ordinance is section 16 of the New Orleans Plumbing Code (See Appendices).

More recently, EPA has introduced guidelines for the pretreatment of pollutants introduced into publicly owned treatment works. These guidelines are presented in the Appendices. Also involved with pretreatment is the user charge system and information on various materials that inhibit biological treatment processes. Information on both of these is presented in the Appendices.

The steps toward the reduction of wastes into sewerage systems by pretreatment has been hastened by EPA guidelines indicating typical waste characteristics (especially for the non-compatible wastes) for various industrial categories.

Also included in these documents are the unit operations or systems that could be applied to correct the problem

Table 3. Corrective Measures Under Emergency Operations Conditions [6].

TREATMENT SYSTEM CATEGORY	EMERGENCY	RESPONSE
Collection System	Line obstruction	1. Analyze situation to determine proper course of action. 2. Implement prevention measures as required — particularly mutual aid agreements. 3. Dispatch a pretrained crew properly equipped. 4. Always check spare parts inventory. 5. Provide portable lighting if at night. 6. Be prepared to cope with traffic. 7. Pump flow around trouble area utilizing portable pumps and quick coupling pipe. 8. Be prepared to cope with sewage backup into nearby buildings, especially those with basements. 9. Restore condition to normal as rapidly as possible. 10. Always clean up the area and treat with lime if spillage occurred. 11. Critique response plan.
Pumping Stations	Equipment failure	1. Analyze the situation to determine the proper course of action. 2. Implement prevention measures as required — particularly mutual aid agreements. 3. Dispatch a pretrained crew properly equipped. 4. Provide portable lighting if at night. 5. Pump flow around trouble area utilizing portable pumps. 6. Critique response plan.
	Equipment failure	1. Check spare parts inventory. 2. Use original equipment quality replacement parts. 3. Always use appropriate lifting and hoist equipment.

TREATMENT SYSTEM CATEGORY	EMERGENCY	RESPONSE
Pumping Stations (Continued)	Equipment failure (continued)	4. Check impellers for blockage due to rags and trash. 5. Check for bearing seizure due to overheating or insufficient lubrication. 6. Check for loose couplings. 7. Always lubricate before restart.
	Power loss	1. Determine if the power loss is local or area-wide. 2. If loss is local, check out all electrical circuits for shorts or system overload. 3. If the loss is area-wide, contact the power company and coordinate repair and start-up operations with them.
	Explosion	1. Determine immediately the cause of the explosion and take action to prevent additional explosions. 2. Notify the fire and police departments and the rescue squads.
Pretreatment	Clogged screens	1. Bypass the unit until the units are operating again. 2. Check cutter for dull blades and replace as required. 3. Check the capacity to be certain units are not hydraulically overloaded. 4. Manually keep the screens clean until problem is corrected. 5. Critique response plan.
Primary treatment	Stoppage of sludge collection mechanism	1. Analyze the situation to determine if repair can be made without draining the tank. 2. Check to see if rags and debris have entwined around the sludge collector mechanism. 3. Check tank bottom for excessive deposits of sand, rock, and other inorganic material.

TREATMENT SYSTEM CATEGORY	EMERGENCY	RESPONSE
Primary treatment (Continued)	Stoppage of sludge collection mechanism (Continued)	4. Stir media manually to lessen or remove accumulations. 5. If applicable, check all drives, chains, and sprockets for malfunctions. 6. Check out the electrical circuit for shorts and system overload. 7. Critique response plan.
Secondary treatment	Rapid sludge removal system malfunctioning	1. Analyze the situation to determine if repair can be made without draining the tank. 2. Open and adjust all suction ports to obtain optimum sludge removal. 3. Backwash system to eliminate clogged condition. 4. Critique response plan.
	Clogged diffuser tubes or clogged sprayer in aeration equipment	1. Replace the clogged unit as soon as possible. 2. Clean the clogged unit immediately upon removal. 3. Check air lines for dirt and trapped water or ice. 4. Critique response plan.
	Clogging and ponding of trickling filter media	1. Check the size of the filter media for nonuniformity. Replace as required. 2. Check for cementing or breaking up of media. 3. Check for fibers, slime growths, trash, insect larvae, or snails in the filter media voids. 4. Stir media manually to lessen or remove accumulations. 5. Flood the filter media for about 24 hours to loosen surface accumulations. 6. Dry growth by drying filter for several hours, if possible. 7. Jet spray areas in filter media with a high pressure water spray. 8. Critique response plan.

Table 4. Wastewater Characteristics — Glass, Cement, Lime, Concrete Products, Asbestos and Gypsum Products. [7]

Characteristic	Flat Glass	Mirrors	Cement & Lime	Concrete Products	Asbestos & Gypsum
Industry Operation	Year-round	Year-round	Year-round	Year-round	Year-round
Flow	Continuous	Continuous	Continuous	Continuous	Continuous
BOD	Low	Low	Low	Low	Low
TSS	Low-HIGH	HIGH	HIGH	Low-HIGH	Low-HIGH
TDS	Low-HIGH	HIGH	HIGH	LOW-HIGH	No data
COD	Low	Low	Low	Low	Low
Grit	PRESENT	PRESENT	Absent	PRESENT	Absent
Cyanide	Absent	Absent	Absent	Absent	Absent
Chlorine Demand	No Data	No Data	No Data	No Data	No Data
pH	Neutral	ALKALINE	ALKALINE	ALKALINE	Neutral
Color	Absent	Absent	Absent	Absent	Absent
Turbidity	Present	Present	Present	Present	Present
Explosive Chemicals	Absent	Absent	Absent	Absent	Absent
Dissolved Gases	Absent	Absent	Absent	Absent	Absent
Detergents	PRESENT	PRESENT	PRESENT	PRESENT	PRESENT
Foaming	Absent	Absent	Absent	Absent	Absent
Heavy Metals	Absent	Present	Absent	Absent	Absent
Colloidal Solids	PRESENT	PRESENT	PRESENT	PRESENT	PRESENT
Volatile Organics	Absent	Absent	Absent	Absent	Absent
Pesticides	Absent	Absent	Absent	Absent	Absent
Phosphorus	PRESENT	PRESENT	DEFICIENT	DEFICIENT	DEFICIENT
Nitrogen	DEFICIENT	DEFICIENT	DEFICIENT	DEFICIENT	DEFICIENT
Temperature	Normal[2]	Normal[2]	Normal-High[2]	Normal[2]	Normal-High[2]
Phenol	Absent	Absent	Absent	Absent	Absent
Sulfide	Absent	Absent	Absent	Absent	Absent
Oil and Grease	ABSENT[1]	Absent	Absent	Absent	Absent
Coliform (Total)	Absent	Absent	Absent	Absent	Absent

[1] Present in automotive glass and glass bottles manufacturing wastewaters.

[2] Temperature equal to or greater than domestic wastewater; may affect design but not harmful to joint treatment.

NOTE: Characteristics which may require pretreatment or are significant to joint treatment plant design are shown in UPPER CASE.

Table 5. Pretreatment Unit Operations for the Glass, Cement, Lime, Concrete Products, Asbestos and Gypsum Products Industry [7].

Pretreatment Group	Suspended Biological System	Fixed Biological System	Independent Physical Chemical System
Flat Glass	grit removal + solids separation	grit removal + solids separation	grit removal + solids separation
Mirrors	grit removal + neutralization + chemical precipitation (heavy metals)	grit removal + neutralization + chemical precipitation (heavy metals)	grit removal + neutralization + chemical precipitation (heavy metals)
Cement & Lime	neutralization	neutralization	neutralization
Concrete Products	grit removal + neutralization	grit removal + neutralization	grit removal + neutralization
Asbestos & Gypsum Products	solids separation	solids separation	solids separation

BIBLIOGRAPHY

1. *Manual of Instruction for Sewage Treatment Plant Operators,* New York State Department of Environmental Conservation.
2. NPDES Permit No. FL0027529, EPA Region IV, Atlanta, Georgia.
3. State of Florida Rules of the Department of Pollution Control, Chapter 17–4, Permits (Supplement No. 35).
4. *Emergency Planning for Municipal Wastewater Treatment Facilities,* EPA-430/9-74-013, USEPA, Office of Water Program Operation, Washington DC 20460, February 1974.
5. *Recommended Standards for Sewage Works,* Great Lakes-Upper Mississippi River Board of Sanitary Engineers, 1971 Revised edition, Health Education Service, P.O. Box 7283, Albany NY 12224.
6. Federal Guidelines, *Operation & Maintenance of Wastewater Treatment Facilities,* USEPA, Office of Water & Hazardous Materials, Washington DC 20460, August 1974.
7. Federal Guidelines, *Pretreatment of Pollutants Introduced into Publicly Owned Treatment Works,* USEPA. Office of Water Program Operations, Washington DC 20460, October 1973.
8. *User Charges and Industrial Cost Recovery,* Denver SMSA, EPA Region VIII, Contract No. 68-01-1864, January 1974.

APPENDICES

A. Allowable Discharges into Sewage Treatment Systems — Section 16, Sewage & Water Board Plumbing Code, City of New Orleans LA.
B. Federal Guidelines — Pretreatment of Pollutants Introduced into Publicly Owned Treatment Works.
C. Prerequisites to a user charge system.
D. Information on Materials Which Inhibit Biological Treatment Processes.

SECTION 16

RULES GOVERNING DISCHARGES INTO THE PUBLIC STORM DRAINAGE SYSTEM AND THE PUBLIC SANITARY SEWERAGE SYSTEM

16.1 RULES GOVERNING DISCHARGES INTO THE PUBLIC STORM DRAINAGE SYSTEM.

(a) GENERAL

(1) The public storm drainage system exists primarily to allow the removal of storm water runoff from public and private land surfaces, and will be referred to herein as the "storm drainage system". The control of pollution in said system and thus the receiving stream which ultimately receives storm drainage water is manifestly in the public interest.

(2) Liquid waste, such as industrial waste, cooling water discharge, water runoff from land surfaces other than resulting from precipitation, etc., may be admitted to the storm drainage system provided they are free of offensive or objectionable pollutants within the limits hereinafter defined as applicable to discharges into the storm drainage system.

(b) ALLOWABLE DISCHARGES: STORM DRAINAGE SYSTEM

Liquid wastes may be admitted to the drainage system, upon security permit, provided:

(1) The wastes do not exceed a B.O.D. loading of 25 pounds per day at a maximum concentration of 15 mg/l.

(2) The wastes do not exceed a C.O.D. loading of 50 pounds per day at a maximum concentration of 30 mg/l.

(3) The wastes do not contain suspended solids in excess of 42 pounds per day at a maximum concentration of 25 mg/l.

(4) The wastes do not give off offensive odors.

(5) The wastes impart no significant coloration to the waters contained in the drainage system.

(6) The wastes do not cause an objectionable change in the pH of the waters contained in the drainage system.

(7) The wastes do not contain materials which form objectionable coatings on the sides or deposits on the bottoms of the drainage systems.

(8) The wastes do not exceed 110 degrees Fahrenheit or raise the temperature of the receiving canal by more than five (5°) degrees Fahrenheit.

(9) The wastes do not contain pathogenic bacteria or the indicator organisms of pathogenic bacteria in quantities greater than the densities prescribed by other agencies as the maximum limit for safe recreational contact waters.

100

(10) The wastes do not contain materials whose concentrations exceed the values in the following list, the analytical results being expressed in terms of the indicated element.

	mg/1
Antimony	0.1
Arsenic	0.05
Barium	1.0
Beryllium	1.0
Bismuth	1.0
Boron	0.5
Cadmium	0.1
Chromium (Hexavalent)	0.05
Chromium (Trivalent)	0.05
Cobalt	0.2
Copper	0.5
Iron	1.0
Lead	0.1
Manganese	0.05
Mercury	0.01
Molybdenum	0.1
Nickel	0.5
Rhenium	0.5
Selenium	0.05
Silver	0.05
Strontium	0.5
Tellurium	0.5
Tin	1.0
Zinc	1.0

(11) The wastes do not contain cyanides or cyanogen compounds in excess of 1.0 mg/l as CN^- in the discharge waste.

(12) The wastes do not contain radioactive materials exceeding the existing standards of the proper regulatory authority.

(13) The wastes do not contain phenols or other taste or odor producing substances in such concentrations as to affect the taste or odor of the receiving stream.

(14) The wastes do not contain unusual concentrations of total dissolved solids (such as sodium chloride or sodium sulfate).

(15) The wastes do not contain hydrogen sulfide in excess of 1.0 mg/l as sulfide ion.

(16) The wastes do not contain any phosphorus as ortho-phosphate ion in excess of 1.0 mg/l.

(17) The wastes do not contain any foaming or frothing agents of a persistent nature.

(18) The dissolved oxygen of the waste is not less than 50% of saturation at the particular discharge temperature of the waste.

(19) The wastes do not contain inorganic nitrogen in the form of nitrite or nitrate ion in excess of 5.0 mg/l.

(20) The wastes do not contain any substance considered or found to be toxic to aquatic life, such as, but not limited to, hydrocarbons, etc.

(c) PROVISIONAL DISCHARGES: STORM DRAINAGE SYSTEM

Liquid wastes which require pretreatment in order to attain the limits required for admission of the discharge to the storm drainage system will be allowed provided there is full compliance with the requirements of Section 16.3 — Rules Governing the Pretreatment of Liquid Waste Discharge into the Public Storm Drainage System and the Public Sanitary Sewerage System.

It shall be expressly understood that the permit for a provisional discharge may be revoked, and the permitted connection to the storm drainage system terminated, at any time the General Superintendent adjudges that such revocation is necessary to protect the receiving streams of the storm drainage system.

16.2 RULES GOVERNING DISCHARGES INTO THE PUBLIC SANITARY SEWERAGE SYSTEM.

(a) GENERAL

(1) The public sanitary sewerage system exists to provide for and allow the collection and/or removal of polluted liquids wastes from public and private property. It is in the public interest that reasonable rules and regulations be applied to discharges into the sanitary sewerage system so as to prevent the system from being; (1) unnecessarily burdened, or (2) excessively burdened.

(2) It is in the public interest that sanitary sewage be treated to remove pollutants, to the degree established by those agencies having jurisdiction, prior to discharge into a receiving stream.

(3) The rules and regulations contained in this Section 16.2 relating to discharges into the sanitary sewerage system of Orleans Parish are supplemental to all other rules and regulations of the Sewerage and Water Board which govern said sanitary sewerage system and tie-ins thereto.

(4) In the event such tie-in is not practicable, industry shall conform to regulations for waste disposal as set forth by the Louisiana State Department of Health and the City Board of Health.

(b) PROHIBITED DISCHARGES: SANITARY SEWERAGE SYSTEM

The discharge of any of the following liquid wastes into the public sanitary sewerage system is prohibited:

(1) Any storm water, surface water, ground water, roof run-off, sub-surface drainage, cooling water or unpolluted industrial process water. These waters shall be discharged into the public storm drainage system, as they would constitute an unnecessary burden upon the sanitary sewerage system.

(2) Any liquid or vapor having a temperature greater than 140 (140°) degrees Fahrenheit.

(3) Any water or wastes which contain wax, grease or oil, plastic or other substance that will solidify or become discernably viscous at temperature between 50 degrees and 90 degrees Fahrenheit.

(4) Any gasoline, benzene, naptha, fuel oil or other flammable or explosive liquid, solid or gas.

(5) Any liquid wastes containing garbage that has not been properly shredded.

(6) Any liquid wastes containing solid or viscous substances in quantities adjudged by the General Superintendent to be capable of causing obstruction or retardation to flow in sewers, or other interference with the proper operation of the sewerage collection and/or treatment system, such as, but not limited to, ashes, cinders, sand, mud, straw, shavings, metal, glass, rags, feathers, tar, plastics, wood, whole blood, paunch manure, hair and fleshings, entrails, lime slurry, lime residue, slops, chemical residues, paint residues, fiberglass, bulk solids, pulped or shredded paper, etc.

(7) Any waters or wastes containing noxious or malodorous substance which can form a gas, which either singularly or by interaction with other wastes, is capable of causing objectionable odors or hazard to life and property, which forms solids in concentrations exceeding limits established herein or creates any other condition deleterious to structures or treatment processes; or requires unusual facilities, attention or expense to handle such materials.

(8) Any waters or wastes containing a toxic and/or poisonous substance in sufficient quantity to injure or interfere with the existing sewage treatment process in sufficient quantity to constitute a hazard to humans or animals, or in sufficient quantity to create any hazard in the receiving streams of the treatment process effluent or proposed future treatment processes; all as adjudged by the General Superintendent.

(9) Any waters or wastes containing free or emulsified oil and grease exceeding, on analysis, an average of 100 mg/l and a maximum of 250 mg/l of either or both combinations of free or emulsified oil and grease.

(10) Any waters or wastes containing free or emulsified oil and grease when, in the opinion of the General Superintendent, it appears probably that such wastes:

(a) Can deposit oil or grease in the sewer lines in such manner as to clog the sewers or impede the flow.

(b) Can overload the sewage treatment facilities skimming and grease handling equipment.

(c) Are not amendable to biological oxidation and will therefore pass to the receiving stream without being affected by the normal sewage treatment processes.

(d) Can have deleterious effects on the sewage treatment process due to excessive quantities or concentrations.

(11) Any waters or wastes which attack or corrode sewers and sewage disposal equipment.

(12) Any waters or wastes having a pH higher than 10.0 or lower than 5.5.

(13) Any waters or wastes containing heavy metals or salts of the heavy metals, in solution or suspension, in concentrations exceeding the following, the analytical results being expressed in terms of the indicated element:

	mg/1
Antimony	0.1
Arsenic	0.05
Barium	3.0
Beryllium	1.0
Bismuth	1.0
Boron	0.5
Cadmium	0.1
Chromium (Hexovalent)	1.0
Chromium (Trivalent)	1.0
Cobalt	0.2
Copper	0.5
Iron	5.0
Lead	0.1
Manganese	1.0
Mercury	0.1
Molybdenium	0.2
Nickel	0.5
Rhenium	0.5
Selenium	0.1
Silver	1.0
Strontium	0.5
Tellurium	0.5
Tin	1.0
Zinc	1.0

(14) Any waters or wastes containing cyanides or cyanogen compounds capable of liberating hydrocyanic acid gas or acidification in excess of one (1) mg/l as CN in the discharged waters or wastes.

(15) Any waters or wastes containing radioactive materials exceeding the existing standards of the Louisiana Board of Nuclear Energy.

(16) Any waters or wastes containing phenols or other taste or odor producing substances in such concentrations as to affect the taste and odor of the receiving

stream after passage through the sewage treatment process.

(17) Any waters or wastes containing unusual concentrations of solids, either suspended or dissolved; as for example, in total suspended solids of inert nature (such as Fuller's Earth) and/or in total dissolved solids (such as sodium chloride or sodium sulfate).

(18) Any waters or wastes causing excessive discoloration not readily removable by the normal sewage treatment process.

(19) Any waters or wastes with excessive B.O.D. or an immediate dissolved oxygen demand.

(20) Any waters or wastes with excessive C.O.D.

(21) Any waters or wastes with excessive hydrogen sulfide concentration.

(22) Any waters or wastes with excessive flow and concentration of any substance resulting in excessive loading of the sewerage system.

(23) Substances which are not amenable to treatment or reduction by the wastewater treatment process employed, or are amenable to treatment only to such degree that the sewage treatment plant effluent cannot meet the requirements of other agencies having jurisdiction over discharge to the receiving stream.

(c) PROVISIONAL DISCHARGES: SANITARY SEWERAGE SYSTEM

Liquid waters or wastes, having a B.O.D. greater than 200 mg/1, having suspended solids greater than 250 mg/1, or having a C.O.D. greater than 400 mg/1, or having combinations thereof, may be allowed discharge into the sanitary sewerage system provided;

Payment is rendered (where applicable) in accordance with the "excessive strength surcharge formula", and provided; The waste is proven and continues to prove amenable to treatment by the particular treatment process which will serve the waste. Any waste that requires pretreatment in order to attain the limits for admission to the public sanitary sewerage system will be considered a provisional discharge.

It shall be expressly understood that the permit for a provisional discharge may be revoked, and the permitted connection to the sanitary sewer terminated, at any time the General Superintendent adjudges that such revocation is necessary to protect the sewage treatment process.

(d) EXCESSIVE STRENGTH SURCHARGES

Any water or waste discharge greater than 10,000 gallons per day, shall be subject to an "Excessive Strength Surcharge" computed on excessive B.O.D. and excessive Suspended Solids by the following formula:

$$S = Vs \times 8.34 \ (\$0.01285 \ (BOD{-}350) + \$0.01208 \ (SS{-}))$$

S = Surcharge in Dollars for the time period applicable.

Vs = Sewage volume in million gallons for the
time period applicable.

$$8.34 = \text{pounds per gallon of water.}$$
$$\$0.01285 = \text{Unit charge for B.O.D. in Dollars per/lb.}$$
$$\text{B.O.D.} = \text{B.O.D. strength index in milligrams per liter}$$
$$\text{when greater than } 350.$$
$$350 = \text{Allowed B.O.D. strength in milligrams per liter.}$$
$$\$0.01208 = \text{Unit charge for SS in Dollars per/lb.}$$
$$\text{SS} = \text{Suspended solids strength index in milligrams per}$$
$$\text{liter, when greater than } 400.$$
$$400 = \text{Allowed suspended solids strength in milligrams per liter.}$$

All assessed surcharges are due and payable on the due date of the normal sewerage service charge and payment of either charge will not be accepted without concurrent payment of the other charge.

(16.3) RULES GOVERNING THE PRETREATMENT OF LIQUID WASTE DISCHARGES INTO THE PUBLIC STORM DRAINAGE SYSTEM AND THE PUBLIC SANITARY SEWERAGE SYSTEM.

(a) GENERAL

The pretreatment of liquid wastes to attain the limits for admission to either the public storm drainage system or the public sanitary sewerage system (as the case may be), will be allowed.

(b) DILUTION NOT ACCEPTABLE

The alteration of the characteristics of a polluted liquid waste, to attain the limits for admission to either the public sanitary sewerage system or the public storm drainage system (as the case may be), by means of dilution will not be allowed as an acceptable pretreatment process. The objective of an acceptable pretreatment process shall be the removal of the pollutant from the liquid waste.

(c) SUBMISSION OF PLANS, SPECIFICATIONS AND DATA OF PRETREATMENT PROCESS.

In the event pretreatment of waters or wastes is required, then all plans, specifications and any other pertinent information relating to proposed treatment, processing facilities or flow equalization facilities, etc., shall be submitted for approval of the General Superintendent prior to the start of construction, if the effluent from such facilities is to be discharged into the public sanitary sewerage system or the public storm drainage system (as the case may be). All such plans shall be prepared by a Registered Professional Engineer and shall bear his signature and seal.

106

(16.4) ·RULES GOVERNING PERMITS FOR DISCHARGES INTO THE PUBLIC STORM DRAINAGE SYSTEM OR THE PUBLIC SANITARY SEWERAGE SYSTEM.

(a) GENERAL

A connection permit must be obtained from the Sewerage and Water Board for any connection to the public storm drainage system or the public sanitary sewerage system (See Section 15).

(b) INDUSTRIAL WASTE CONNECTION PERMIT

(1) In addition to the regular connection permit required by Section 15, any person, partnership or corporation desiring discharge of an industrial waste (defined as "The liquid wastes from any industrial manufacturing process, trade or business, as distinct from sanitary sewerage.") or a combination of industrial waste with sanitary sewage, shall secure a specific "Industrial Waste Connection Permit" prior to discharge into the public storm drainage system or the public sanitary sewerage system (as the case may be).

(2) Permission for connection to the public sanitary sewerage system or the public storm drainage system (as the case may be), will not be granted until industry has filed application furnishing the General Superintendent with an analysis characterizing their waste. The analytical procedures, where applicable, shall follow those procedures set forth in *Standard Methods for the Examination of Water and Wastewater,* APHA, AWWA, WPCF, latest edition. The application shall also include pertinent information relating to average flows, peak flows, average loading, peak loadings, etc. Applications shall be filed with the House Connection Department and will be processed as expeditiously as possible

(c) EXISTING INDUSTRIAL WASTE CONNECTIONS

Existing connections to the public storm drainage system or the public sanitary sewerage system which would be classified under the provisions of this Section 16 as "Industrial Waste Connections" are not exempt from securing the required permits, and must file application for said permit.

(d) PENALTY FOR VIOLATION OF RULES

The General Superintendent shall lay before the Special Counsel of the Board any cases of the violation of these or other rules that may be herein provided, and the Special Counsel shall cause the proper charges to be made and vigorously prosecute the offenders in such cases to the full extent of the law.

(e) POWERS AND AUTHORITIES OF ENFORCEMENT

The General Superintendent or his duly authorized agent bearing identification and credentials shall be permitted to gain access to such properties as may be necessary for the purpose of inspection, observation, measurement, sampling and testing, to determine compliance to the provisions of this Code and should a violation of this Code be round, the industry shall be served by the General Superintendent with written notice stating the nature of such violation and providing a time limit for the satisfactory correction thereof. Any industry who shall continue any violation beyond the prescribed time limit shall be in violation of this Code and shall be authority for the Sewerage and Water Board to shut off the water supply to the offending premises and terminate the connection receiving discharge from the offending premises. Any industry in violation of this Code shall become liable to the Sewerage and Water Board for any expense, loss or damage sustained by the Sewerage and Water Board by reason of such violation.

**Pretreatment of Pollutants Introduced
Into Publicly Owned Treatment Works [7]**

SECTION I

INTRODUCTION

1. *Purpose*

These guidelines are established to assist municipalities, States, and Federal agencies in developing requirements for the pretreatment of wastewaters which are discharged to publicly owned treatment works. The Guidelines also explain the relationship between pretreatment and the effluent limitations for a publicly owned treatment works.

The U.S. Environmental Protection Agency (EPA) has published Pretreatment Standards in 40 CFR 128 (Appendix A). The standards will be enforceable by the EPA. These guidelines provide technical information useful to States and municipalities in establishing pretreatment requirements to supplement the Federal pretreatment standards.

2. *Authority*

Authority for these guidelines is contained in Section 304(f) (1) of the Federal Water Pollution Control Act Amendments of 1972 (the ACT), which states:

> "For the purpose of assisting States in carrying out programs under Section 402 of this Act, the Administrator shall publish . . . guidelines for pretreatment of pollutants which he determines are not susceptible to treatment by publicly owned treatment works. Guidelines under this subsection shall be established to control and prevent the discharge . . . of any pollutant which interferes with, passes through, or otherwise is incompatible with such works".

3. *Definitions*

Compatible Pollutant:

Biochemical oxygen demand, suspended solids, pH, and fecal coliform bacteria, plus additional pollutants identified in the NPDES permit if the publicly owned treatment works was designed to treat such pollutants, and in fact does remove such pollutants to a substantial degree. The term substantial degree is not subject to precise definition, but generally contemplates removals in the order of 80 percent or greater. Minor incidental removals in the order of 10 to 30 percent are not considered substantial. Examples of the additional pollutants which may be considered compatible include:

> Chemical oxygen demand
> Total organic carbon
> Phosphorus and phosphorus compounds
> Nitrogen and nitrogen compounds
> Fats, oils, and greases of animal or vegetable
> origin (except as prohibited where these materials
> would interfere with the operation of the publicly
> owned treatment works).

Incompatible pollutant:

Any pollutant which is not defined as a compatible pollutant.

Joint Treatment Works:

Treatment works for both non-industrial and industrial wastewater.

Major Contributing Industry:

A major contributing industry is one that: 1. has a flow of **50,000** gallons or more per average work day; 2. has a flow greater than five percent of the flow carried by the municipal system receiving the waste; 3. has in its waste a toxic pollutant in toxic amounts as defined in standards issued under Section 307 (a) of the Act; or 4. has a significant impact, either singly or in combination with other contributing industries, on a publicly owned treatment works or on the quality of effluent from that treatment works.

Pretreatment:

Treatment of wastewaters from sources before introduction into the joint treatment works.

4. *The Federal Water Pollution Control Act Amendments of 1972*

The Act established a national system for preventing, reducing, and eventually eliminating water pollution. The ultimate goal is to eliminate the discharge of pollutants into the navigable waters of the United States.

Under the National Pollutant Discharge Elimination System (NPDES), all point sources (including publicly owned treatment works) must obtain a permit for the discharge of wastewaters to the navigable waters of the United States.

The Act further requires that, as a minimum intermediate objective, all point sources other than publicly owned treatment works treat their wastewaters by the application of the best practicable control technology. Subsequently, the minimum requirement for industrial wastewater would be the application of best available treatment technology. For publicly owned treatment works, the initial objective is secondary treatment, followed by best practicable treatment technology.

Monitoring for compliance with effluent limitations and pretreatment requirements will be in accordance with EPA guidelines established under Section 304 and implemented under Section 308 of the Act.

SECTION II

EFFLUENT LIMITATIONS NPDES PERMITS

5. *NPDES Municipal Permits*

Procedures developed under Section 402 of the Act will provide the details for implementation of permit programs. The purpose of the following discussion is to highlight the relation bebetween the permit programs and the pretreatment standards and guidelines.

Under the National Pollutant Discharge Elimination System (NPDES), all point sources (including publicly owned treatment works) must obtain a permit for the discharge of wastewaters to the navigable waters of the United States. Permits will not be required for industrial sources discharging into publicly owned treatment works.

The pretreatment standards (Appendix A) will be directly enforceable by EPA on industry.

The effluent limitations for a pollutant in the discharge from a publicly owned treatment works will be individually determined by the permitting agency, based on information supplied by the municipality. These effluent limitations will be included in the discharge permit issued to the municipality for the publicly owned treatment works.

Additionally the permit for a publicly owned treatment works will require provisions for adequate notice to the permitting agency of:

a. New introductions into such works of pollutants from any source which would be a new source as defined in Section 306 of the Act if such source were discharging pollutants.

b. New introductions of pollutants into such works from a source which would be subject to Section 301 of the Act if it were discharging such pollutants.

c. A substantial change in volume or character of pollutants being introduced into such works by a source already discharging pollutants into such works at the time the permit is issued.

This notice will include information on the quantity and quality of the wastewater introduced by the new source into the publicly owned treatment works, and on any anticipated impact on the effluent discharged from such works.

111

The permit programs developed under Section 402 of the Act should be consulted for details regarding permit application

6. *Effluent Limitations for Publicly Owned Treatment Works*

There are various sources of effluent limitations, including:

a. Effluent limitations for publicly owned treatment works (Section 301(b) of the Act). Secondary treatment information is contained in 40 CFR Part 133 (Appendix B).
b. Toxic Effluent Standards or Prohibitions (Section 307 (a) of the Act).
c. Water Quality Standards (Section 303 of the Act).

The most stringent limitation for each pollutant will govern.

SECTION III

JOINT TREATMENT AND PRETREATMENT

7. *Joint Treatment*

Joint treatment of industrial and municipal wastewaters in the same treatment works is generally a desirable practice. Treatment of the combined wastewaters can benefit the environment, the municipality, and industry if properly designed and operated.

Some of the advantages of joint treatment are:

a. Increased flow which can result in reduced ratios of peak to average flows.
b. Savings in capital and operating expenses due to the economics of large-scale treatment facilities.
c. Better use of manpower and land.
d. Improved operation (larger plants are potentially better operated than smaller plants).
e. Increased number of treatment modules with resultant gains in reliability and flexibility.
f. More efficient disposal of sludges resulting from treatment of wastewaters containing compatible pollutants.

In some cases, the characteristics of the industrial wastewaters may be beneficial in the publicly owned treatment works processes. For example, some industrial wastewaters contain organic material but are devoid of the nutrients required for biological treatment. In joint biological treatment the nutrients are present in the domestic wastewater and consequently do not have to be added (as they would in separate industrial treatment). For a plant required to remove nutrients, the joint biological treatment of nutrient-free organic industrial wastes would result in lower amounts of nutrients to be removed in a subsequent process.

There may be characteristics present, however, which make a wastewater not susceptible to joint treatment if introduced directly into the municipal system. In such cases, it will be necessary to pretreat the wastewater.

8. *Pretreatment Policy*

The following are basic pretreatment policy considerations used in developing these guidelines:

a. Joint treatment of domestic wastewaters and adequately pretreated industrial wastewaters is encouraged where it is the economical choice.
b. In-plant measures to reduce the quantity and strength of industrial wastewater flows can be beneficial to joint treatment, and should be encouraged.
c. Pretreatment for removal of compatible pollutants is not required by the Federal pretreatment standards.
d. In recognition of the broad spectrum of industries, waste constituents, and treatment plants, State and municipal pretreatment requirements should be based on an individual analysis of the permitted effluent limitations placed on a publicly owned treatment works and on the potential for adverse effects on such works.

9. *Federal Pretreatment Standards*

EPA has issued standards in 40 CFR Part 128 for pretreatment of pollutants introduced into publicly owned treatment works (Appendix A). These standards are designed to protect the operation of a publicly owned treatment works and to prevent the introduction of pollutants into publicly owned treatment works which would pass through such works inadequately treated.

The pretreatment standards are intended to be national in scope, i.e. generally applicable to all situations. In many cases, it will be necessary for a State or a municipality to supplement the Federal Standards with additional pretreatment requirements which take into account local conditions. The purpose of these guidelines is to provide information to States and municipalities to assist in the development of these supplemental pretreatment requirements. The following paragraphs describe the basis and extent of the Federal pretreatment standards.

Section 128.131 of the Standards (Appendix A) is designed to protect the operation of publicly owned treatment works. Wastes which are prohibited from introduction into publicly owned treatment works are listed. This Section is applicable to non-domestic users only.

Section 128.133 is designed to prevent the introduction of incompatible pollutants to publicly owned treatment works which would pass through such works inadequately treated. Any pollutant that is not a compatible pollutant as defined in Section 128.121 is by definition incompatible. Section 128.133 is applicable only to "major contributing industries" as defined. Pretreatment is required for incompatible pollutants to the levels of best practicable control technology currently available as defined for industry categories in guidelines issued pursuant to Section

304(b) of the Act. Provision is made for the Administrator to segment the industrial users of municipal systems as a special category for the purposes of defining best practicable control technology currently available. Provision is also made to permit a less stringent pretreatment standard for an incompatible pollutant if the municipality is committed in its NPDES permit to remove a specified percentage of the incompatible pollutants introduced into a publicly owned treatment works generally should not pass through such works in amounts greater than would be permitted for direct discharge.

Biochemical oxygen demand, suspended solids, pH, and fecal coliform bacteria, which are defined as compatible, are the pollutants used to describe the effluent quality attainable by secondary treatment in 40 CFR Part 133 (Appendix B). Not later than July 1, 1977, secondary treatment effluent limitations must be met by all publicly owned treatment works which discharge into navigable waters unless more stringent effluent limitations are necessary to ensure compliance with water quality standards or toxic effluent standards.

The terms compatible and incompatible pollutant should not be misinterpreted. For incompatible pollutants, it will be necessary for the municipality to assess the capabilities of its treatment works in order to determine whether it can make a commitment in its NPDES permit to remove some percentage of the incompatible pollutants in the treatment works. Such a commitment might be made of there is some removal of incompatible pollutants in the treatment works which occurs as an incident to the removal of compatible pollutants. In this case it must be determined that the incidental removal can be relied upon and will not cause harm to the treatment works or interfere with its operation (including sludge handling and disposal processes).

There will also be situations when the Federal pretreatment standards (without credit for removal in the publicly owned treatment works) will not be sufficient to protect the operation of the publicly owned treatment works. This might be the case when the quantity of an incompatible pollutant introduced by all major contributing industries would result in a concentration of the pollutant in the influent to the treatment works which would inhibit the performance of the treatment process. In such a case, the municipality would have to supplement the Federal standards.

Pretreatment for removal of compatible pollutants is not required by the Federal pretreatment standards. This is based on the premise that pretreatment facilities should not be required for removal of compatible pollutants as a substitute for adequate municipal waste treatment works. This, however, does not preclude State or local pretreatment requirements for compatible pollutants. Pretreatment of wastewaters containing compatible pollutants may be necessary in the form of spill protection or flow equalization in order to ensure compliance with the Federal pretreatment standards and permitted effluent limitations.

PREREQUISITES TO A USER CHARGE SYSTEM [8]

The selection and adoption of a user charge system is not a simple matter. Many diverse factors and considerations influence the formulation of user charges.

Public Acceptance

Not surprisingly, the concept of user charges and industrial cost recovery promulgated by EPA is not universally embraced or even understood. Attitudes concerning user charges may well include the view that it is an "insidious form of taxation," or "in conflict with economic development activities," or "inconsistent with economic principles" (decreasing unit costs). Further, it may be viewed as "dictatorial" or "federal interference in local affairs." All of these views have been reflected in various newspaper accounts from localities across the nation in recent months.

Without debating the relative merits of user charges versus other forms of revenue measures, suffice it to say that a municipality or other agency contemplating the adoption of a schedule of user charges to meet EPA guidelines is advised to conduct an enlightened program of citizen education, provide strong policy leadership, and insure that its proposals are based upon factual, technical information.

Industrial Data

It is simply not possible to develop an equitable and realistic program of industrial user charges without data as to the requirements which industrial users place upon the wastewater system. Typical base data requirements include:
1. Type establishment; principal products
2. Number of employees by work shift
3. Source(s) of water supply
4. Volume of water use
5. Uses of water
6. Type processes utilized
7. Waste disposal points
8. Composition, volume and flow rate of wastes discharged

Such data will permit the wastewater agency to identify the potential discharges of non-domestic wastes, as well as the characteristic magnitude and composition of those discharges. This facilitates the selection of those industries to be included in the sampling program.

To the wastewater agency, a "know your industry" posture is essential to effective wastewater management. Regular personal contacts should supplement technical data collection and sampling programs.

One of the prerequisites to approval of construction grants is a determination by EPA that the applicant "has legal, institutional, managerial, and financial capability to insure adequate construction, operation and maintenance of treatment works."

It is not beyond comprehension that an agency failing to develop or employ adequate base data on users and treatment costs may be found to not meet this criterion.

Accounting And Cost Allocation

The preceding study, *Institutional and Financial Arrangements for Wastewater Management — Denver SMSA,* revealed a wide disparity of accounting practices and a general failure (except among such agencies as MDSDD #1 and City and County of Denver) to maintain a classification of accounts which permitted cost allocation to the various wastewater cost components. This reveals a general variance from guideline requirements that adequate documentation for user charge measures be maintained.

To develop a user charge system to meet EPA guidelines appears to dictate that a separate wastewater "fund" be established within the accounting systems of all agencies; further, that this fund be also broken down into functional account codes such as "administration," "collection system," "pumping," "primary treatment," "slidge disposal," etc. Sub-classifications within these functions should include "salaries and wages," "maintenance and repair," "capital expenditures," "debt service," etc. A purely "line item" classification will not produce the necessary information.

Obviously, particularly in the smaller agencies, it is difficult to allocate the time of a work crew or piece of equipment which, in the course of a day or week performs in several functional areas. Yet, it should be possible to develop a relatively simple (if necessary, subsidiary) accounting procedure to allocate costs to the various functions as needed to support and document the user charge derivation. Without this, there is no realistic or equitable way to allocate costs to function or to users.

User Charge Schedule

While the EPA guidelines define "treatment works" to include all physical components of the wastewater system, there are certain capital cost components which may or may not be recovered by user charges. Primary among these are the collection system and the local share of a capital grant program.

The EPA guidelines neither require not prohibit financing the initial installaation or subsequent extension of the collection system from user charges. Most wastewater agencies regard the construction of the collection system as of unique special benefit to the property served, and levy a tap charge, frontage charge, or special assessment for connection which is sufficient to recovery of the costs of the extension. For those agencies which operate under this principle, the user charges are then designed only to provide maintenance and repair of the collection system, plus the other cost components of the total wastewater system. The derivation and use of the "plant investment fee," as currently levied by some agencies in the SMSA, may be subject to scrutiny under the established guidelines unless it is based on a "user charge" type allocation to reflect relative system usage.

The local share of a capital grant improvement program may be recovered at the descretion of the local agency. If it is a system-wide improvement affecting all users, equitibility could be achieved by recovery through the user charge system. If it is of special benefit to only a segment of system customers (i.e., a secondary collection system) more equitibility could be achieved by recovery through some type of special assessment to the affected customers.

In some instances in the Denver SMSA, the initial construction of subsequent extension of the collection system is financed by the mill levy. So long as the cost or value of the extension is reflected in enhancement of the value of the property served, this would appear to be equitable and consistent with EPA guidelines. Should industry be afforded preferential treatment, by comparison with other users, in the amount of cost burden levied for main extensions, it is conceivable that EPA would find this practice inconsistent with the guidelines unless it is demonstrated as the most equitable method of allocating costs. On the other hand, financing of treatment facilities and operations by mill levies would almost certainly be inconsistent with the letter and the intent of these guidelines.

The findings of this study do little to support the application of the SIC classifications to either users or user classes, except as perhaps an interim or transitional measure. Wide variances in waste characteristics among users of the same class are reported by wastewater agencies maintaining sampling/monitoring programs.

Volume, biochemical oxygen demand (BOD) and suspended solids (SS) are the typical measures employed as rate factors in the user charge schedule. In particular circumstances, in particular localities, dissolved solids (DS), chemical oxygen demand (COD), nutrients and other paramters (color, pH, deleterious substances, etc.) may be covered by a rate factor or by standards and prohibitions governing discharges.

Adopting Ordinance

The EPA guidelines require that the user charge schedule be "incorporated in one or more municipal legislative enactments or other appropriate authority," or "ordinances" in conventional terminology. While sanitation districts may not have been granted authority to enact ordinances *per se*, it is believed that their statutory authority to fix schedules of rates and charges is sufficient to enable these agencies to do so in a manner to satisfy the intent of the guidelines.

In the course of the study, several agencies were identified which fix user charges by negotiated contract with the individual industry, rather than by ordinance. This practice should be examined in the light of the guidelines, both as to form and to content, to insure compliance. It should also be examined in the light of state constitutional or legislative mandates that practices be uniform and equitable among classes.

Of course, no agency is required to enact user charges in the form specified, unless it desires to be eligible to receive federal construction grant funds. The guidelines do require however, that all municipalities or districts which are part of a regional system must have enacted such charges for the regional system to become eligible. The implications of this requirement are discussed elsewhere in this report.

The adopting ordinance, in addition to the user charge schedule, should also establish standards and requirements for waste discharges (See Appendix A for Model Ordinance) and other features of the agency's industrial waste program.

Operating and Management Plan

It is not believed to be in the overall public interest or in the interests of effective wastewater management that the wastewater systems become merely "receivers" of wastes, levying such charges as may be appropriate under circumstances. Rather, it is believed that each agency should develop an active operating and management plan or strategy for dealing with its industrial users.

For example, the agency should work with its industrial users to alleviate peaking conditions, either hydraulic or chemical, which create added operational costs. Industries might be prevailed upon to cooperatively agree to discharge at off-peak times, or on a staggered schedule to minimize operational and cost requirements.

Working realtionships between the wastewater agency and the individual industries should be such that changes in industrial processes or other changes in demands upon the wastewater system would be made known to the wastewater agency well in advance so that appropriate adjustments could be made.

Both the user charge schedule and the operating plan (as well as the water charge schedule) should encourage pre-treatment and other reductions in waste volumes and concentrations with which the wastewater system must deal. The "know your industry" posture, previously cited, should facilitate a partnership understanding.

Included with the operating and management plan must be an orderly, systematic program for the monitoring of industrial wastes. This program must include physical on-site provisions for gathering samples, laboratory testing and analysis of those samples, a realistic schedule for taking samples, a continuing program of information and data exchange with industrial users, and some "early warning" mechanism to avoid disruption or overloading of treatment plant processes.

Physical provisions may include a requirement for sampling manholes, and may also include a requirement that the industry provide holding tanks for temporary deteration and gradual release of strong wastes. Many agencies vary the frequency of gathering samples in accordancee with the strength of the wastes; the greater the strength, the greater the frequency of sampling.

Plan Update

The user charge and industrial cost recovery requirements of the EPA guidelines stipulate periodic (at least annual) update of the user charge schedule to reflect changed conditions.

Ideally, a continuous review and adjustment would be accomplished. However, stability in rate structures (just as in tax structures) is vital to the economic stability and well-being of commercial and industrial users. To avoid widely-fluctuating rate structures from year to year, the wastewater agency should develop enlightened capital improvement programs and financial policies which anticipate outlays in an orderly fashion, consistent with need but also consistent with revenue considerations.

Transitional Steps

The smaller agencies, in particular, will find it impractical to instantly initiate an industrial waste program. Physical and financial constraints may render this impossible. The guidelines would appear to anticipate and permit this, so long as a definite program of implementation is underway.

For example, until physical arrangements have been made for a sampling program, the wastewater agency might require that industrial users secure laboratory analyses of their wastewater from independent contract agencies, using the lab report as a basis for charges. Alternatively, agencies might jointly staff and equip the necessary industrial waste program, or contract with a larger agency for its performance.

Until accounting classification and cost allocation procedures have been developed, estimates may have to be made of the various cost components of the system. Until individual characteristics of users have been determined, SIC or other classifications may have to be applied.

Finally, until the program and its impact have been fully developed, a deferred effective date might be established. During this period, samplings can be taken, industries can evaluate and provide pre-treatment measures, and cost accounting techniques can be perfected. "Sample" bills might be sent to industrial users during this period. In other localities, marked reductions in loadings have been accomplished when deferred dates or "warning periods" are established. The end result would appear to be completely consistent with national goals and with EPA guidelines and regulations.

INFORMATION ON MATERIALS WHICH INHIBIT BIOLOGICAL TREATMENT PROCESSES [7]

The following paragraphs comprise a discussion of the sources of information on various materials, which inhibit biological treatment systems, presented in this Appendix.

Copper

Rudolfs has indicated that copper in raw sewage at 1 mg/L either inhibited or retarded the aerobic metabolism. Barth, et al and McDermott, et al have shown, from pilot-plant studies, that continuous addition of 1.0 mg/L of copper is the threshold value for activated sludge. The maximum concentration of copper that can be received continuously without having a detectable effect on common parameters for effluent quality is 1.0 mg/l. Where there is concern about turbidity, a concentration of 0.8 mg/L of copper appears to be the upper limit beyond which an effluent on pilot-plant studies conducted with copper sulfate and copper-cyanide complex additions.

When four-hour slug doses of copper were added to the activated sludge system, the performance was affected when the copper concentration in the feed approached 75 mg/l. The same studies have indicated that heavy metals present in sewage (whether alone or in combination) affected the nitrification process and that no acclimation was possible. Kalabina, et al have indicated that copper fed as copper sulfate inhibited nitrification when the concentration of copper exceeded 0.5 mg/L. They also recommended a copper concentration of less than 0.1 mg/L in raw sewage for biological treatment.

With respect to anaerobic digestion, a copper concentration of 10 mg/L in raw sewage affected gas production in the digestion of primary sludge. When anaerobic digestion of combined primary and secondary sludges was involved, a copper concentration of 5 mg/L in raw sewage affected the digestion process. However, other reports indicated that copper concentrations in the range of 0.2 mg/L to 2.5 mg/L in raw sewage were deleterious to the anaerobic digestion process.

Zinc

McDermott, et al and Barth, et al, have shown from pilot-plant studies that zinc (either in the form of $ZnSO_4$ or in the form found in a typical alkaline cyanide plating bath) added to raw sewage produced similar effects on biological systems. The maximum level of zinc that will not produce a significant effect on treatment efficiency was indicated as being between 2.5 and 10 mg/L when continuously added to the biological system.

When zinc at a concentration of 160 mg/L was added as a slug dose over a four-hour period, a serious reduction in treatment efficiency was observed.. The same authors have shown that the anaerobic digestion process was retarded when the zinc concentration in raw sewage reached 20 mg/1. This concentration is in agreement with those reported by Rudolfs. Other investigators have indicated a maximum concentration of 5 mg/L Zn in raw sewage to prevent decrease in gas production in the anaerobic digestion process.

Zinc concentrations of approximately 0.5 mg/L in raw sewage have been reported to inhibit nitrification process.

Lead

Rudolfs has indicated that lead in raw sewage at concentrations of about 0.1

mg/L will inhibit or retard the aerobic biological metabolism. This is in agreement with the values reported by Kalabina from studies conducted using lead sulfate.

Kalabina also studied the effects of lead on nitrification process and showed that 0.5 mg/l of lead in raw sewage inhibited the nitrification process.

Cadmium

Mosey reported that cadmium ions, added to experimental digesters, have a threshold value (for adverse effect on biologic treatment processes) of 0.02 mg/l and occurs at a carbonate ion concentration of 2×10^{-5} moles/l. The results also showed that the toxicity of cadmium to anaerobic digestion was pH-dependent when the pH was greater than 7.0 and independent of the pH when the pH was less than 7.0.

The foregoing results were based on laboratory studies, and the cadmium concentrations refer only to those present in the digester.

Boron and Arsenic

Banerji, et al studied the effects of boron added in slug doses to activated sludge treatment. The results indicated that a slug dosage of 10 mg/L in raw waste adversely affected the performance of aerobic metabolism. However, Rudolfs indicated that boron in raw sewage affected the performance of activated sludge and trickling filters at much lower levels 91.0 mg/L and less). He also indicated that the addition of arsenic at concentrations of approximately 4 mg/L to digesting sludge inhibited or retarded the performance of digesters.

Chromium

Rudolfs indicated that chromium was toxic to activated sludge and trickling filter processes when the raw sewage contained 3 mg/L total chromium. It was pointed out that when the total chromium in raw sewage was in the range of 1–5 mg/L, the anaerobic digestion process was significantly retarded.

However, the pilot-plant results of Barth, et al indicate that a continuous dosage of 10 mg/l of sludge pilot-plant was able to withstand up to 500 mg/L of chromate chromium when applied over a four-hour period. Their studies also showed that the anaerobic digestion process was affected when the hexavalent chromium in raw sewage exceeded 50 mg/L. These values were significantly higher than those reported by Rudolfs. According to Rudolfs hexavalent chromium concentrations at 1.0 mg/L in raw sewage affected trickling filter process, and at the 1–50 mg/l level the anaerobic digestion process was affected.

It is important to point out here that the toxicity of heavy metals on biological systems is closely associated with sulfate concentrations in raw wastewater and therefore should not be compared directly with one another without considering the concentration of the sulfate ion. Unfortunately, the studies did not report the sulfate concentrations necessary for making such a comparison.

With respect to the nitrification process, Whiteland, et al have reported that the nitrification process was severely affected when the hexavalent chromium in raw sewage was in the range of 2 to 5 mg/L.

Nickel

Pilot-plant studies conducted with continuous addition of $NiSO_4$ in raw sewage indicated that nickel concentrations at 2.5, 5, and 10 mg/L affected both biological treatment efficiency and effluent clarity. The results indicated that nickel doses of 1 mg/L on a continuous basis can be tolerated by aerobic biological processes. However, it is important to point out that most of the nickel reaching the aeration process passed through the effluent in soluble form. These results also indicated that the anaerobic digestion process was very resistant to nickel in the sludges. Primary sludges containing up to 40 mg/L of nickel digested satisfactorily.

However, other reports indicate that nickel concentrations of 2 mg/L in raw sewage would be inhibitory to the anaerobic digestion process. In addition, the nitrification process will be severely inhibited when the nickel concentration is approximately 0.5 mg/L.

Cyanides

Coburn reported that 5 mg/L of cyanide in raw wastewater (discharging continuously) interfered with activated sludge treatment. In trickling filter studies, cyanide in raw sewage at 30 mg/L produced poor effluent quality. However, when the cyanide concentration was only 10 mg/L, 98 to 100 percent of the cyanide was destroyed in the trickling filter. These levels are higher than those reported by Rudolfs. According to Rudolfs, 1 to 2.0 mg/L of cyanide (as HCN) in raw sewage affected the performance of the activated sludge, trickling filters, and anaerobic digestion processes. Generally, however, secondary biological treatment processes can oxidize the cyanide if acclimatized. Rudolfs has also reported that the nitrification process was inhibited when the cyanide (HCN) concentration in raw sewage reached 2 mg/L.

Sulfides and Sulfates

Pohland and Kang reported in their review of literature on anaerobic processes that when the sulfate concentration exceeded 500 mg/L, the gas production was greatly reduced. Similar results were also reported indicating that sulfate concentrations of 300 mg/L are toxic to anaerobic digestion. The sulfate toxicity was related to the reduction to sulfides during digestion.

Lawrence, et al and Rudolfs have reported that sulfide concentrations in the range of 150–200 mg/L in the digester feed would reduce gasification considerably.

Ammonia

Drague, et al reported that ammonia concentrations in the range of 1,500–3,000

122

mg/L inhibited the anaerobic digestion process. When the ammonia concentration reached 3,000 mg/L, the feed sludge became strongly toxic to the digestion process, this is in agreement with the values reported by Rudolfs. The concentration of ammonia refers to that present in the influent to the digesters of accumulated in the digesters.

Sodium Chloride

The presence of sodium chloride in raw sewage or in the digester has been reported to produce deleterious effects both in aerobic biological systems and in anaeroc digesters. Rudolfs has indicated that the gas production in anaerobic digestion will be greatly reduced when the NaC1 concentration reaches 50,000 mg/L in digesters. Studies conducted by Kincannon indicated that the laboratory-scale aerobic biological system was greatly affected at NaC1 concentrations of 10,000 mg/L in raw wastewater. Similar results were reported by Lawton and Eggert in the case of trickling filters.

Chloroform

Ghosh, in a review of the anaerobic digestion process literature, reported that chloroform addition to digesters on a continuous basis produced noticeable effects at a concentration of 10 to 11 mg/L. When chloroform application was as a slug dosage, the anaerobic process was adversely affected at 1 mg/L concentration in the feed.

Free Oil

Free oil of petroleum origin, at concentrations in the range of 50–100 mg/L, has been reported as interfering with aerobic biological treatment. The oil concentrations were measured according to the API Manual using CCl_4 extraction. The measurement of free oil using the API procedure may also include oily materials of animal and vegetable origin.

Other Cations (Na^+, K^+, Ca^+, and Mg^{++})

Malina has presented data on the effects of several cations on anaerobic digestion. The results indicated that: Ca^{++} strongly inhibited anaerobic digestion at a concentration of 8,000 mg/L; Mg^{++} at a concentration of 2,000 mg/L; K^+ at 12,000 mg/L; and Na^+ at 8,000 mg/L. These results were in close agreement with those reported by Kugelman and McCarty.

Chlorinated Organic Compounds

The effect of several chlorinated organic compounds, at very low concentrations, on anaerobic digestion of biological sludges is marked, and constitutes a warning that a close watch must be kept on the disposal of these compounds. Jackson, et al have summarized the results on toxicity of various chlorinated organic compounds, as indicated in Table 1. These results indicate that chlorinated compounds can be toxic to anaerobic digestion at levels varying between 0.1 to 20 mg/L. The degree of inhibition will depend on the particular chemical itself. Further studies are required to determine the long-term effects of chlorinated organic compounds on municipal biological treatment processes when they are present separately or in various combinations.

Table 1. Information on Materials Which Inhibit Biological Treatment Processes.

Pollutant	Concentration[1], mg/L		
	Aerobic Processes	Anaerobic Digestion	Nitrification
Copper	1.0	1.0	0.5
Zinc	5.0	5.0	0.5
Chromium (Hexavalent)	2.0	5.0	2.0
Chromium (Trivalent)	2.0	2000^2	*
Total Chromium	5.0	5.0	*
Nickel	1.0	2.0	0.5
Lead	0.1	*	0.5
Boron	1.0	*	*
Cadmium	*	0.02^2	*
Silver	0.03	*	*
Vanadium	10	*	*
Sulfides (S$^-$)	*	100^2	*
Sulfates (SO$_4^-$)	*	500	*
Ammonia	*	1500^2	*
Sodium (Na$^-$)	*	3500	*
Potassium (K$^-$)	*	2500	*
Calcium (Ca^{--})	*	2500	*
Magnesium (Mg^{--})	*	1000	*
Acrylonitrite	*	5.0^2	*
Benzene	*	50^2	*
Carbon Tetrachloride	*	10^2	*
Chloroform	18.0	0.1^2	*
Methylene Chloride	*	1.0	*
Peintachlorophenol	*	0.4^2	*
1,1,1-Trichloroethane	*	1.0^2	*
Trichlorofluoromethane	*	0.7	*
Trichlorotrifluoroethane	*	5.0^2	*
Cyanide (HCN)	*	1.0	2.0
Total oil (Petroleum origin)	50	50	50

* Insufficient data, [1] Concentrations refer to those present in raw wastewater unless otherwise indicated. [2] Concentrations apply to the digester influent only. Lower values may be required for protection of other treatment process units. [3] Petroleum-based oil concentration measured according to the API Method 733-58 for determing volatile and non-volatile oily materials. The inhibitory level does not apply to oil of direct animal or vegetable origin.

CHAPTER 6

COMMON OPERATING PROBLEMS AND SUGGESTED SOLUTIONS*

INTRODUCTION

There is a variety of common operating problems which may occur periodically and prevent the proper processing of wastes by a treatment plant.

The purpose of this section is to properly identify the problem by defining the indicators. Once the problem has been identified, certain monitoring, analyses and/or inspections must be performed prior to making a decision as to which corrective measures should be utilized. In some cases, the data-gathering process can be a simple visual observation and in other cases it can involve rather intricate sampling and laboratory procedures. The resulting information should then be systematically utilized to make a determination on which of the corrective measures should be implemented.

The problems discussed in this section are those which occur rather frequently in practice and the suggested solutions have been, for the most part, accepted procedure in the industry. There may be times when the suggested corrective measures do not correct the problem or a problem may exist which does not fit into the commom category. In a case such as this, it is prudent to seek expert advice on the subject prior to undertaking any course of action. The information utilized in development of the indicators and solutions reflect the present state-of-the-art.

---

*Procedures for Evaluating Performance of Wastewater Treatment Plants, Contract No. 68-01-0107, USEPA, Office of Water Programs.

VI. METERING

VII. SOLIDS HANDLING

I. PRETREATMENT — Pumping Plants and Influent Sewers

Problem — SURGING OF PLANT INFLUENT

Indicators

1. Intermittent flooding of weirs and structures.
2. Plant efficiency treating wastewater drops sharply for short period of time.
3. Flow meter records intermittent high and low peak flows.
4. Excessive suspended solids in overflows.

Monitoring, Analysis and/or Inspection

1. If a main sewage lift station pumps effluent to plant, check for frequent starting and stopping of pumps or more than one pump operating at one time (out of phase) during a pumping cycle.
2. If influent flows to the plant by gravity through a main trunk, check depth of flow in connection sewers if channel is uniform or monitor flow with a portable flow meter.
3. If surging occurs during rainfall, record relation of surging to duration of rainfall; and record or obtain rainfall intensity if possible.

Corrective Measures

2. Surging from gravity influent line indicates a major pump station discharge into a connecting sewer. Review water depth and/or portable flow meter data to determine source of flow.
1. Intermittent starting and stopping or recycling of main influent pump station or stations indicates improper wet well sensor adjustment or that the hydraulic capacity of the station has been exceeded. Adjust level sensors for a more desirable pumping cycle. If possible, install variable speed pumps units for uniform flow into treatment plant; or install surge tank.
3. Heavy surging or hydraulic loading of treatment plant treating waste from a separated system during periods of normal rainfall indicates illegal connections to system such as catch basins, yard drains, or roof downspouts. "Smoke bomb" sanitary sewer system to determine source of illegal connections.
4. Heavy surging or hydraulic loading of treatment plant during period of heavy rainfall is caused primarily by flooding of street areas and water entering the system through manholes, broken lines, etc. Seal all manhole covers in high risk flood areas and patch all cracks in manhole structures with an epoxy water resistant compound.

Indicators

1. Scum blanket in wet well
2. Odors
3. Improper operation of level sensing equipment

Monitoring, Analysis and/or Inspection

1. Sound wet well with a pole to determine solids level.
2. Measure wet well draw down during pumping cycle.
3. Ascertain relation of pump suction piping to floor of wet well.
4. Determine elevation of inlet piping.
5. Look for dead spots in corners and structural cracks where sludge can accumulate.

Corrective Measures

1. Start pumps manually, being careful not to break suction, and pump wet well down to lowest possible elevation while breaking scum blanket with a high pressure water hose.
2. Check pumping level and determine if more of a drawdown can be allowed for in order to remove more of the floatable materials.
3. If one or more influent lines come in at a higher elevation than the pump suction inlet, set level sensor so drawdown will allow for spillage of fresh wastewater onto the scum blanket. The resultant turbulence could assist in breaking the blanket.
4. If this is a persistent problem install air diffusers in wet well with compressors wired to operate in tandem with the pumps. Diffused air will assist in placing the solids in suspension and curtail the development of a scum blanket.

Problem — ODOR SOURCE IN WET WELL

Indicators

1. Odors of hydrogen sulfide origin
2. Corrosion if iron work and concrete
3. Black color observed in liquid or solids

Monitoring, Analysis and/or Inspection

1. Hang hydrogen sulfide (lead acetate) indicator tiles in wet well.
2. Sample wastewater in wet well and analyze for total and dissolved sulfides.
3. Check for floating solids in wet well.
4. Run dye test on influent sewer or sewers to determine velocity of waste flow and travel time to wet well.
5. Check temperature of wastewater in wet well.
6. Check pump invert position and condition.
7. Check passage time in the interceptors at flow velocities.

Corrective Measures

1. Low velocities (less than 2 ft/sec) are an indication that solids are being deposited in influent sewer and sulfides are being formed and released in the wet well. Velocity must be increased in the sewer or influent must be continuously treated upstream with chlorine or copper as to prohibit the development of hydrogen sulfide gas.
2. If the source of hydrogen sulfide is in the wet well and not the influent sewer, increase pumping cycle for more frequent removal of solids.
3. Install air diffusers in wet well to keep wastewater fresh.
4. Install blower and gas scrubber for the oxidation of the gases and exhausting to the atmosphere.
5. Dose wet well with hyperchloride on a periodic basis to suppress the formation of hydrogen sulfide.

I. PRETREATMENT — Screening and Shredding

Problem — ACCUMULATION OF RAGS AND DEBRIS FOR DISPOSAL

Indicators

1. Large amount of rags and debris accumulated on plant site gives off obnoxious odors and attracts flies and other insects.

Monitoring, Analysis and/or Inspection

1. Estimate volume (cubic feet) of rags and debris removed each day in proportion to flow.
2. Determine time exposed material is allowed to accumulate.
3. Check disposal method used.

Corrective Measures

1. Arrange for local refuse or garbage company to pick up rags and debris on a daily basis and dispose of them in a sanitary fill.
2. Store rags and debris in closed containers whenever possible.
3. If incineration facilities are available on plant site or at some other convenient location, burn them. Care must be taken to see that the emission from the incinerator meets location air pollution control requirements.
4. Rags and debris can be disposed of on the plant site if sufficient land is available for a fill and cover operation.
5. If none of the above methods proves feasible, rags and debris can be ground up by installing the proper equipment and returned to the plant flow. This method should only be used as a last resort since ground or shredded screenings may cause problems in pumping equipment and in digesters.

Problem — EXCESSIVE GRIT IN BAR SCREEN CHAMBERS

Indicators

1. Surging in chamber due to increase in water level.
2. Low removal of grit by degritting equipment.
3. Excessive dryout of grit with screening.

Monitoring, Analysis and/or Inspection

1. Sound chamber with flat board at end of pole to determine depth of grit.
2. Determine velocity in chamber by timing a dye release from one end of chamber to the other.
3. Check plans and probe bottom of chamber to determine whether there are any irregularities in chamber bottom slope.
4. Check channel when dewatering for regularly scheduled maintenance.

Corrective Measures

1. If velocity in chamber is less than 2 ft/sec. flush chamber regularly with high pressure water hose. If slide gates are available at inlet end of chamber, throttle gates to jet flow along chamber bottom.
2. Remove irregularities or reslope chamber bottom; if possible, to increase velocity.
3. Regulate velocity in grit chamber by utilizing different outlet weir shapes.

Problem — ODOR SOURCE IN GRIT CHAMBER

Indicators

1. Odors of hydrogen sulfide origin
2. Corrosion of metal work and concrete

Monitoring, Analysis and/or Inspection

1. Hang hydrogen sulfide (lead acetate) tiles in chamber.
2. Check velocities through grit chamber.
3. Check volatile solids content of the grit.
4. Sample wastewater in chamber and analyze for total and dissolved sulfides.
5. Check for floating solids in chamber.
6. Measure depth of grit in chamber.
7. Check for submerged rags and debris on bar screen.

Corrective Measures

1. Clean bar screen thoroughly so as not to impede flow.

2. Increase velocity to 1 fps.
3. Wash grit chamber thoroughly daily with high pressure water hose to move sludge and floating solids through screens.
4. Dose chamber with hyperchloride on a periodic basis to suppress the formation of sulfide. Excessive doses of this chemical should be avoided as it can be toxic to biological treatment systems and anaerobic digestion systems.
5. Install blower and gas scrubber for the oxidation of gases and exhausting to the atmosphere.

Problem — SHREDDED SCREENINGS, CLOGGING PUMPS

Indicators

1. Rope-like rags and debris wrapped around pump impellers.
2. Pump suction lines plugged with "bundles" of rags.
3. Excessive pump or drive beating and power requirements.
4. Appearance of chunks larger in size than usual in the shredder discharge.

Monitoring, Analysis and/or Inspection

1. Install pressure gages on discharge side of pump and check pressure daily.
2. If pump discharge line visible, check flow daily.
3. Check power input, bearing heat pressures on inlet and discharge sides.
4. Check driver speed and power train for power use and delivery to the shredder.

Corrective Measures

1. Do not shred screenings and return them to flow. Remove them either mechanically or manually and dispose of them by burial.
2. Check cutters periodically on all barminutors, comminutors, and other shredding equipment for sharpness.
3. Back flush pumps, periodically if possible.
4. If necessary, modify pump type, inlet protection or prior protective devices to avoid recurred plugging.
5. Upgrade screening and grit removal operation.

I. PRETREATMENT — Grit Handling and Removal

Problem — GRIT REMOVED HAS HIGH ORGANIC CONTENT

Indicators

1. Grey color of grit

2. Odors from grit
3. Greasy feel with excessive components

Monitoring, Analysis and/or Inspection

1. Run volatile solids test on grit daily.
2. Check discharge pressure on cyclonic grit removal equipment.
3. Check velocities with dye releases in grit chambers.
4. If grit chamber is aerated, check air flow rate to chamber.
5. Visual examination of grit and identification of origin of grit materials.

Corrective Measures

1. Keep pressure on cyclonic grit removal equipment at an acceptable range (usually between 4 and 6 psi) by governing pump speeds.
2. Increase velocities in grit chambers by whatever means possible.
3. Adjust air accordingly.
4. Check inlet and outlet controls, baffles and mechanical equipment; adjust and repair and keep clean.

I. PRETREATMENT — General

Problem — INDUSTRIAL WASTE IS INADEQUATELY PRETREATED.

Indicators

1. Discoloration of influent
2. Sterilization of biological treatment processes
3. Digester upsets
4. Change in influent odor
5. Unusual amounts of solids in influent

Monitoring, Analysis and/or Inspection

1. Constantly monitor influent for pH above 8.0 or below 6.0.
2. Check pH of raw sludge.
3. Run heavy metal tests on influent.
4. Check influent temperature.
5. Run settleable solids test.
6. If toxic flow is constant, attempt to trace source upstream of treatment plant.
7. Run C.O.D. test on contaminated effluent and compare results with normal plant loading.

Corrective Measures

1. Bypass all biological treatment processes to parallel units as soon as contaminant

has been detected.
2. Isolate and dispose of all contaminate sludges.
3. If digesters show increases in volatile acids because of contaminated sludges, see Section VII(b) for corrective measures.
4. If activated sludge process has become contaminated, dispose of floc and restart process.
5. Institute program of source control (industrial waste ordinances).

II. PRIMARY TREATMENT — Primary Sedimentation Tanks

Problem — FLOATING, GASEOUS, OR SEPTIC SLUDGE IN TANKS

Indicators

1. Floating material in tank deadspots or in scum troughs
2. Odors of hydrogen sulfide origin

Monitoring, Analysis and/or Inspection

1. Run total solids tests on raw sludge being pumped from primary sedimentation tanks with sample being taken at beginning and end of pumping cycle.
2. Dewater tanks and check sludge collector mechanism (flights, chains and scrappers) for wear and tear.
3. Observe conditions of tanks prior to chemical treatment of tank influent.

Corrective Measures

1. If total solids of raw sludge analyzed at end of pumping cycle is over 2%, increase duration of pumping cycle, preferably with a timer.
2. If sludge collector mechanism shows signs of wear during inspection, repair or replace.
3. Certain chemicals, such as alum, used in chemical treatment, cause floating sludge if this chemical is recirculated to the primary sedimentation tanks. Change operating procedures or chemicals used.
4. Long detention in sludge hoppers favors production of a sludge that readily slips.

Problem — LOW SETTLEABLE SOLIDS REMOVAL EFFICIENCY

Indicators

1. Floating and gaseous sludges in tanks
2. Percent settleable solids removal below 95%

Monitoring, Analysis and/or Inspection

1. Run settleable solids test (Imhoff Cone) during times of day where there are

appreciable changes in plant flow.

2. Check raw sludge removal pumping cycles and duration of pumping period.
3. Run total solids test on raw sludge removed from tanks both at beginning and end of pumping cycle.
4. Dismantle and/or inspect raw sludge pumps and sludge collection mechanism for wear and tear.
5. Check tank inlets with relation to the tank outlets. If baffles have been installed on the inlets, dewater the tanks and check their condition.
6. Calculate theoretical detention time, weir overflow rates, and surface loading rates and compare all data with design criteria.
7. Try a dye test to estimate flow-through time. Check for density stratification due to significant temperature or density difference top to bottom.

Corrective Measures

1. If efficiency of removal drops during peak or increased plant flows, the hydraulic capacity of tanks has probably been exceeded. Refer problem to operating agency engineering staff.
2. Repair all worn raw sludge pumps, parts and sludge collector mechanism.
3. Damaged or missing inlet line baffles could cause tank short circuiting whereby increased velcoities from one end of the tank to the outlet end cause settleable matter to remain in suspension. Replace or repair baffles.

Problem — ERRATIC OPERATION OF SLUDGE COLLECTION MECHANISM

Indicators

1. Frequent replacement of broken sheer pins on chain driven collector mechanisms.
2. Frequent torque switch activated alarms on concentric driven clarifier equipment.
3. Visible slippage or "stuttering" of clarifier sludge collection mechanisms.

Monitoring, Analysis and/or Inspection

1. Check all drives for gear wear.
2. Dewater tank and check chains and sprockets for wear and see that chains have not come off sprockets.
3. Check to see that rags and debris have not entwined themselves around sludge collector mechanism.
4. Check dewatered tanks for excessive bottom deposits of sand, rocks and other inorganic material.
5. Sound bottom for excessive accumulation of sludge.

Corrective Measures

1. Repair all worn sludge collector equipment and drives.
2. If rags are a problem, make provisions for removal of all rags and debris as part of the pretreatment process.
3. If sand and rock depsoits on the tank bottom are a problem, provide adequate screening and grit removal as a part of the pretreatment process.
4. If sludge accumulation is a problem, increase frequency of pumping raw sludge from tanks.

Problem — LOW SCUM (GREASE) REMOVAL

Indicators

1. Visible grease particles being discharged in plant effluent
2. Excessive water in scum pits

Monitoring, Analysis and/or Inspection

1. Run grease test (composite) on plant influent and effluent and calcculate efficiency of grease removal equipment and compare with plant design criteria.
2. Observe if wooden flights making a return travel on tank surface carry grease particles adhered to them under scum troughs at the discharge end of the tanks.
3. Determine, with a pole, the depth of floating scum and water in scum pits.
4. Check capacity of scum pits.
5. Check for wear of scum pickup wiper blades.

Corrective Measures

1. If possible, lower return wooden flight to below water surface so grease particles do not adhere to them.
2. Install water sprays to direct grease particles on tank surface into scum troughs. Water spray should not break surface tension on water surface.
3. If scum removal is done manually and intermittently, continuous removal equipment should be installed.
4. Excessive water in scum pits should first be removed by pumping from bottom of pit to plant headworks and then the concentrated scum can be pumped to a digester or an incinerator.
5. Efficiency of scum removal in plants receiving a high grease loading can be increased by the addition of flotation or evacuator equipment.
6. Since grease particles normally in suspension tend to agglomerate into larger particles after being dosed with chlorine, chlorine contact tanks should be provided with grease removal equipment.
7. Pump scum pits down on a regular basis so as not to cause scum overflows back into the clarifier.
8. Clean and replace all worn wiper blades.

Problem — TANK CONTENTS TURN SEPTIC

Indicators

1. Tank contents are a dark color
2. Hydrogen sulfide odors emitted from tanks

Monitoring, Analysis and/or Inspection

1. Run total and dissolved sulfide tests on both tank influent and contents.
2. Run DO test.
3. Check pH of tank influent.
4. Check quantity and total solids of all inflows into tank from other plant processes such as digesters supernatant, thickener overflows, centrifuge concentrates, etc.

Corrective Measures

1. If tank influent contains high dissolved and total sulfides, influent is septic. Prechlorinate influent or correct problem at source.
2. If tank influent pH is below 6 or above 8, toxic waste is being discharged into plant and must be corrected at the source.
3. If discharges from other plant processes contain excessive total solids and exceed 5% of the daily tank inflow, the sedimentation tank is being overloaded. If possible, reduce rate of process flows to sedimentation tanks or pretreat flows by aerating or chlorinating them. If possible divert or find other means of disposal for supernatant or centrates.

III. SECONDARY TREATMENT — Activated Sludge Process

Problem — SLUDGE BULKING

Indicators

1. Rising sludge in final clarifiers
2. Floating matter being discharged in final effluent
3. Filamentous growths in mixed liquor

Monitoring, Analysis and/or Inspection

1. Check mixed liquor for low pH and low dissolved oxygen.
2. Check sludge age, food/micro-organism ratio, or mean cell residence time.
3. Run settleability test and check for separation of floc in graduated cylinder or use Mallory Direct Reading Settleometer.

4. Check aeration period.
5. Check for NO_2 and NO_3.

Corrective Measures

1. If possible, reduce organic loads on aeration tanks affected.
2. Add digested sludge (that has been aerated for some time) to aeration tanks.
3. Dose the aeration tanks with alum or ferric chloride together with lime.
4. Controlled chlorination of return activated sludge.
5. Reduce sludge age and air rate to stop nitrification.
6. Increase sludge age by regulating waste sludge rate.
7. Increase or correct low dissolved oxygen or pH in aeration tank.
8. Increase aeration period by placing another aerator in operation if possible or reduce the return sludge rate by thickening the return sludge concentration by coagulation.
9. Control filamentous growth by increasing sludge age or supplementing nutrient deficiencies.
10. If nitrification is desired, maintain a minimum of 3 mg/1 NO_3 and a maximum of 1 mg/1 NO_2.

Problem — ERRATIC SLUDGE VOLUME INDEXES

Indicators

1. Pin floc visible in final clarifiers overflow
2. Poor settling characteristics of mixed liquor

Monitoring, Analysis and/or Inspection

1. Check mixed liquor suspended solids in each aeration tank.
2. Run 30 minutes settleability test in each aeration tank.
3. Determine whether the point floc is a recurrent situation or the result of toxicants.

Corrective Measures

1. Regulate wasting to decrease suspended solids in mixed liquor.
2. Chlorinate return activated sludge.
3. Decrease solids loading to aeration tanks.
4. Make appropriate adjustments to obtain a less oxidized sludge.

Problem – DIFFICULTY IN MAINTAINING BALANCED MIXED
LIQUOR AND DISSOLVED OXYGEN IN AERATION TANK

Indicators

1.   Intermittent sludge bulking
2.   Loss of sludge blanket in secondary clarifier
3.   Dark color in the aerator contents

Monitoring, Analysis and/or Inspection

1.   Check D.O. concentration in different areas of aeration tanks during changes in daily flow.
2.   Check suspended solids in mixed liquor at different periods during thd day.
3.   Run suspended solids test on aerator influent and mixed liquor to check sludge age.
4.   Monitor rate of flow to aeration tanks.
5.   Check daily flow variation in loading for excessive peak demand periods.

Corrective Measures

1.   Lower D.O. concentrations occurring during changes in plant flow are an indication of excessive loading of aeration tanks. Increase air supply to tank if possible by placing another blower into service, etc.
2.   Decrease loading to aeration tanks by placing more tanks into service if possible.
3.   Provide controlled air supply to aeration tanks by interlocking blower speeds to tank D.O. monitoring equipment.
4.   Increase inflow could hydraulically overload the secondary treatment system. If possible, bypass a portion of the flow from the primary sedimentation tank until the flow rate returns to normal.

Problem – EXCESSIVE FOAM IN AERATION TANKS

Indicators

1.   Frothing in aeration tanks

Monitoring, Analysis and/or Inspection

1.   Check influent for radical temperature changes.
2.   Run M.B.A.S. test on influent.
3.   Check suspended solids concentration in aeration tanks.
4.   Check D.O. concentration in aeration tanks.

Corrective Measures

1. If practical, increase mixed liquor suspended solids by decreasing wasting rate.
2. Install or operate reclaimed water sprays in aeration tanks.
3. Utilize defoaming agent.
4. Lower air supply while being careful to maintain a safe dissolved oxygen concentration in aeration tanks.

Problem — DIGESTER SUPERNATANT AND/OR CENTRIFUGE
CENTRATE UPSETTING ACTIVATED SLUDGE PROCESS

Indicators

1. Mixed liquor changes from a light to a dark brown color
2. Mixed liquor suspended solids decrease sharply during operation of centrifuge or when sludge is being pumped to digester.
3. Final clarifier effluent increases in turbidity
4. Mixed liquid DO decreased.

Monitoring, Analysis and/or Inspection

1. Run total solids and pH tests on supernatant or centrate being returned to plant headworks.
2. Check mixed liquor suspended solids during discharge of centrate or supernatant.
3. Continuously check mixed liquor D.O.
4. Check volume of concentration of recycle flows relative to total plant flow.

Corrective Measures

1. Program discharges of supernatants or centrates so they do not coincide.
2. Pump centrates to digester whenever possible.
3. If centrates have high solids content, use flocculants in the operation of the centrifuge.
4. If supernatant have high solids content, experiment with supernatanting from a different level in the digester.
5. If possible, pre-aerate centrate and supernatants prior to discharging them to the activated sludge process.
6. Avoid discharging supernatants from septic digesters to activated sludge process.
7. If possible, release only small amounts of supernatant or centrate during periods of low infow and increase return activated sludge rates if necessary.
8. Program recycles so that the return load does not exaggerate peak loading.

Problem — UNABLE TO MAINTAIN BALANCED FOOD/MICRO
ORGANISM RATIO IN AERATION UNIT

Indicators

1. Fluctuation in S.V.I.
2. Fluctuation in sludge age.

Monitoring, Analysis and/or Inspection

1. Check S.V.I. at least daily.
2. Check mixed liquor suspended solids at least daily.
3. Check suspended solids in influent and effluent at least daily.
4. Monitor plant flow, return activated sludge rate, and waste activated sludge rate.

Corrective Measures

1. Select and operate secondary treatment system by either Mean Cell Residence Time, Solids Retention Time, food/micro organism ratio, or sludge age.

Problem — FACILITIES INADEQUATE FOR DISPOSAL OF WASTE
ACTIVATED SLUDGE (HIGH SUSPENDED SOLIDS IN THE CLARIFIER
OVERFLOW)

Indicators

1. High suspended solids in mixed liquor
2. Increased turbidity in final clarifier effluent
3. High solids concentration in digester supernatants and sludge thickener effluents

Monitoring, Analysis and/or Inspection

1. Check S.V.I. at least daily.
2. Run C.O.D. and suspended solids of plant influent with and without supernatant and thickener return flows.
3. Check blanket depth level or high and low load periods.
4. Check inlet and outlet baffling for possibility of short circuiting.

Corrective Measures

1. If waste activated sludge is not settling in sludge thickener, increase flow of raw sludge to thickener or dose thickener inflow with coagulants.
2. Attempt to break "sludge cycle" by lowering wasting rate.

3. Re-aerate waste sludge prior to pumping to thickeners or discharging to primary clarifiers.
4. Change operation mode of aeration tanks to contact stabilization if possible.
5. If the sludge doesn't settle satisfactorily, coagulate with iron or aluminum, but adjust pH if necessary to keep aerator pH within 6.0–8.5

Problem — UNEVEN HYDRAULIC AND SOLIDS LOADING OF AERATION TANKS

Indicators

1. Mixed liquor suspended solids in each aeration tank varies considerably.
2. Dissolved oxygen in each tank varies considerably.

Monitoring, Analysis and/or Inspection

1. Run S.V.I. in each tank at least daily.
2. Run mixed liquor suspended solids in each tank at least daily.
3. Run dissolved oxygen in each tank at least daily.
4. Run suspended solids on influent and effluent.

Corrective Measures

1. Adjust valves and inlet gates to equalize flow to all tanks when operating under conventional mode.
2. Equalize air flow to all tanks by throttling valves on air discharge lines.

III. SECONDARY TREATMENT — Trickling Filters

Problem — ICE BUILDUP ON MEDIA

Indicators

1. Visible ice formation on filter media

Monitoring, Analysis and/or Inspection

1. Check air temperature.
2. Check recirculation rate to filter.
3. Check flow through filter orifices.
4. Check temperature of wastewater flow to filter.
5. Check filter surface for even distribution of flow.

Corrective Measures

1. By regulating amount of recirculation rate, adjust flow to filter to prohibit the formation of ice.
2. Adjust flows from orifices and splash plates to reduce spray effects.
3. Cover filter to reduce heat losses or install a windbreak to reduce chill factor.
4. Manually break up and remove major ice formations.
5. If possible, add hot water or steam to filter influent.

Problem — FILTER ODORS

Indicators

1. Odors of hydrogen sulfide origin present
2. Black slime visible on surface of filter media

Monitoring, Analysis and/or Inspection

1. Check dissolved and total sulfide of plant and filter influents.
2. Check filter drains for stoppages or growths.
3. Check rate of recirculation to filter.
4. Check for filter overflow or splashing.

Corrective Measures

1. If flow to filter is septic, correct in upstream system by aeration or controlled prechlorination.
2. Clear under drain system of all stoppages.
3. Force air into filter drain system to increase ventilation through filter media.
4. Increase recirculation rate to filter to increase D.O. and to slough off surface slime.
5. Keep areas around filters clean of slimes and growths.
6. Cover filter with inert material and exhaust air into an odor control/scrubber.

Problem — FLY NUISANCE IN VICINITY OF FILTER

Indicators

1. Tiny gnat sized flies becoming a nuisance in plant area and in neighboring area

Monitoring, Analysis and/or Inspection

1. Inspect grounds for tall grass, weeds and other sanctuaries for filter flies.

Corrective Measures

1. Increase rate of recirculation to filter to wash fly larvae out of filter.
2. If possible, flood filter for approximately 24 hrs. to prevent completion of life cycle of flies.
3. Apply a low dosage of chlorine being careful not to sterilize filter media.
4. Maintain grounds so as not to provide sanctuaries for flies.

Problem — CLOGGING AND PONDING OF FILTER MEDIA

Indicators

1. Ponding on filter surface
2. Intermittent flooding of filter

Monitoring, Analysis and/or Inspection

1. Check size of filter media for uniformity.
2. Check for cementing or breaking up of media.
3. Check for fibers, slime growths, trash, insect larvae, or snails in filter media voids.
4. Check organic loading on filter.
5. Check hydraulic load on filter.

Corrective Measures

1. If filter media is non-uniform and the smaller pieces fill the voids, replace the media.
2. Jet problem areas in filter media with a high pressure water spray from a stationary distributor.
3. Stir media manually to lessen or remove any accumulations.
4. Dose the filter media with chlorine at a rate of 5 mg/1 for several hours a day during periods of low flow.
5. Flood filter media for approximately 24 hours to loosen surface accumulations.
6. Dry growth by drying filter for several hours, if possible.

Problem — CLOGGING OF DISTRIBUTOR NOZZLES CAUSES UNEVEN DISTRIBUTION OF FLOW ON THE FILTER SURFACE.

Indicators

1. Uneven sprays from distributor nozzles
2. Ponding on certain areas of the filter media with concurrent drying of other areas

Monitoring, Analysis and/or Inspection

1. Attempt to identify types of solids clogging nozzles.
2. Check for visible grease particles in waste being pumped to filter.
3. Run settleable solids test of waste being pumped to filter.

Corrective Measures

1. Remove and clean all nozzles and thoroughly flush distributor piping.
2. Improve primary clarifier skimming to prevent grease carryover to filter.
3. Increase detention time in primary tanks to prevent settleable and suspended solids carryover to filter.

III. SECONDARY TREATMENT — Oxidation Ponds

Problem — EXCESSIVE WEEDS AND TULES

Indicators

1. Excessive weed and tule growths
2. Mosquito problems in neighborhood of ponds
3. Poor pond circulation

Monitoring, Analysis and/or Inspection

1. Check water depths in selected areas of the pond.

Corrective Measures

1. Deepen all pond areas shallower than three feet.
2. Remove all weed and tule growths as soon as they are visible.
3. For mosquito control, vary liquid level in the pond every 10 days.

Problem — POND ODORS

Indicators

1. Odors of hydrogen sulfide origin from pond
2. Other objectionable odors

Monitoring, Analysis and/or Inspection

1. Check for blue-green algae growths in pond.
2. Check for scum accumulation in pond.
3. Analyze for total and dissolved sulfides in pond and pond influent.

4. Check pond pH and pond influent pH.
5. Check DO content in pond at several locations.

Corrective Measures

1. If pond influent is septic, correct situation upstream by aeration or controlled prechlorination.
2. If possible, aerate pond with mechanical aerators.
3. Remove or break up all scum accumulations.
4. Prechlorinate pond influent.
5. If pond is septic, divert flow from aerobic pond to it or pump high D.O. make up water to it.
6. Add sodium nitrate to pond.
7. Provide odor masking agent if feasible.

Problem — LOW POND DISSOLVED OXYGEN

Indicators

1. Low algae growth in pond
2. Trace hydrogen sulfide odors
3. Grey color of pond

Monitoring, Analysis and/or Inspection

1. Check all areas in pond for adequate D.O.
2. Monitor flow into pond and calculate average daily detention time in pond.
3. Check pH of pond influent and pond contents.
4. Run total and dissolved sulfides in pond influent.
5. Check pond loading rate (lb BOD/acre).
6. Check for floating aquatic weeds

Corrective Measures

1. Increase detention time in ponds to at least five days by placing ponds in parallel.
2. In the abscence of adequate D.O. in the pond, aerate pond contents or pond influent.
3. Chlorinate pond influent if sulfides are present.
4. Physically remove floating weeds to increase light penetration.

III. SECONDARY TREATMENT — Final Sedimentation

Problem — SLUDGE OR PIN FLOC FLOWING OVER WEIRS

Indicators

1. Particulate material rising to surface in clarifier
2. Effluent clarity poor

Monitoring, Analysis and/or Inspection

1. Check clarity of water in final clarifier with Secchi disc.
2. Measure turbidity of effluent discharged from clarifier with turbidimeter.
3. Attempt to determine height of sludge blanket in clarifier with depth sampler and/or a wooden pole.
4. Run suspended solids test on final clarifier effluent.
5. Check all sludge uptake piping to see that they are flowing freely and that 60% of the sludge removed is from inner 50% area of the clarifier.
6. Check pump rates and schedules of sludge withdrawal pumps.
7. Dewater clarifiers and check for damage on sludge scraper mechanism especially at the periphery of the tank.
8. Determine if the weir overflow rate is equal for the entire weir.

Corrective Measures

1. Increase pumping rate for greater removal of sludge from clarifier.
2. Wash down or clean all sludge uptake piping.
3. Repair or replace all damaged sludge scraper mechanism.
4. Level the weir.
5. If uneven weir overflow rate is caused from wind, install a windbreak.
6. Persistence of this problem would indicate a malfunction in the secondary treatment process.

Problem — ERRATIC OPERATION OF SCRAPER MECHANISM

Indicators

1. Frequent torque switch activated alarms on concentric driven clarifier equipment
2. Visible slippage or "stuttering" of clarifier sludge collection mechanisms

Monitoring, Analysis and/or Inspection

1. Check all drives for gear wear.
2. Dewater tanks and check for true travel of scrapper mechanism.

Corrective Measures

1. Repair all worn sludge collector equipment and drives.

ADVANCED TREATMENT — Chemical Coagulation and Flocculation

Problem — CHEMICAL COAGULANTS UTILIZED FOR SETTLING OR DEWATERING SLUDGES THROUGH RECYCLING CAUSE FLOATING SLUDGES IN PRIMARY SEDIMENTATION TANKS

Indicators

1. Floating and gaseous sludge floating in primary sedimentation tanks during periods of chemical treatment
2. Poor settling characteristics in the sludge
3. Poor sludge dewatering characteristics

Monitoring, Analysis and/or Inspection

1. Determine types and amounts of chemicals used.
2. Conduct laboratory jar tests to determine effect of recycled chemicals on waste-water.
3. Check the dosing sequence, the rapid mix energy, and flocculation energy.

Corrective Measures

1. Correct the dosage of coagulants and alkalinity in line with process requirements on the basis of phosphorus content and a coagulant metal/phosphorus ratio of about 2/1. Institute a regular phosphorus determination and product turbidity control.
2. Correct coagulant concentrations and dose points to favor efficient chemical usage.
3. Make sure that sufficient rapid mix energy (800–1000 G) is applied for 1/2 to 2 minutes after dosage to mix substrate and coagulant. Follow with lower energy flocculation to agglomerate fines.
4. Adjust coagulant dosage in line with flow and concentration variations in the plant inflow. Monitor the treated overflow turbidity by the hour.

IV. ADVANCED TREATMENT — Ammonia Stripping

Problem — FREEZING IN AMMONIA STRIPPING TOWER

Indicators

1. Ice formation on outside face of tower

2. Drop in ammonia removal efficiency

Monitoring, Analysis and/or Inspection

1. Monitor ammonia removal
2. Check out water temperature for maximum and minimum values.
3. Check air circulation rate and distribution.

Corrective Measures

1. Use large flow distribution orifices at the outside face of the tower thus concentrating a curtain of warm water where the cold air first enters the tower.
2. Reverse draft fan to blow warm inside air outward to melt the ice.
3. If ammonia removal efficiency drops below 30%, take the tower out of operation.

Problem — FORMATION OF CALCIUM CARBONATE SCALE ON AMMONIA STRIPPING TOWER FILL AND STRUCTURAL MEMBERS

Indicators

1. Visible scale deposits in tower

Monitoring, Analysis and/or Inspection

1. Check pH of tower influent (ph has to be 10.5 or greater.)

Corrective Measures

1. Hose off scale with water jet at periodic intervals.
2. Install water sprays in tower for jetting off scale at frequent intervals.
3. Clean tower with light solution of sulfuric acid.

IV. ADVANCED TREATMENT — Filters

Problem — FILTER BACKWASH WASH WATER HYDRAULICALLY OVERLOADS AND UPSETS CLARIFIERS

Indicators

1. Surging of clarifier during filter backwash operation
2. Higher turbidity in the treated flow

Monitoring, Analysis and/or Inspection

1. Determine amount and rate of filter backwash water being recycled to clarifiers.
2. Increase drum speed, backwash pressure or temperature, and backwash rate.

Corrective Measures

1. Collect backwash wastes in a storage tank and recycle at a controlled rate to clarifiers.

IV. ADVANCED TREATMENT — Microscreen

Problem — FOULING OF FABRIC WITH GREASE AND SOLIDS

Indicators

1. Loss of microscreen efficiency
2. Visible solids and grease on fabric

Monitoring, Analysis and/or Inspection

1. Run suspended solids test on influent to screen.
2. Check for upset in activated sludge process, if any.

Corrective Measures

1. Increase speed of drum.
2. Increase backwashing pressure.
3. Adjust backwashing cycles.
4. Backwash with hot water.
5. Backwash with an approved degreasing agent.

Problem — MICROSCREEN EFFLUENT EXHIBITS HIGHER SUSPENDED SOLIDS CONTENT THAN INFLUENT.

Indicators

1. Increased suspended solids and turbidity on discharge side of screens.

Monitoring, Analysis and/or Inspection

1. Determine colloidal solids in the influent to microscreens.
2. Determine total solids removal efficiency of microscreens.
3. Determine C.O.D. removal efficiency of microscreen.

Corrective Measures

1. Convert colloids in microscreen influent to suspended solids by adjusting the pH accordingly or by the addition of coagulants.

IV. ADVANCED TREATMENT — Activated Carbon

Problem — MECHANICAL FOULING OF COLUMNS

Indicators

1. The flushing of accumulated solids from carbon columns is followed by an abrupt increase in the rate of absorption
2. Pressure rises in downflow columns
3. Decrease in flow rate.

Monitoring, Analysis and/or Inspection

1. Run suspended solids and total organic carbon of column influent and effluent.
2. Check for upset in secondary treatment process.
3. Check column operating routine and backwash records.

Corrective Measures

1. Bypass column influent with large amounts of suspended matter.
2. Flush columns frequently.
3. Consider upflow operation, filtration, or larger carbon grain size.

V. DISINFECTION — Chlorination

Problem — INSUFFICIENT CHLORINE GAS PRESSURE AT THE CHLORINATOR WITH ALL CYLINDERS CONNECTED TO GAS PHASE

Indicators

1. Chlorine pressure gage at chlorinator is reading too low.
2. Chlorine supply lines from cylinders are either very cold or are icing.
3. Chlorine cylinders or cylinder show a frost line

Monitoring, Analysis and/or Inspection

1. Reduce feed rate on chlorinator to about one-tenth the rotameter capacity.
2. If after a short period the chlorine gas pressure rises appreciably it can be concluded that the rate of feed through the chlorinator is greater than the

evaporation rate of the chlorine cylinders or cylinder at the prevailing ambient temperature.

Corrective Measures

1. Connect enough cylinders to the supply system so that the chlorine feed rate does not exceed the withdrawal rate of the cylinders. For 150 lb. cylinders the withdrawal rate at room temperature is 40 lbs. per day per cylinder; for ton containers it is 400 lbs. per day. At lower temperature it is less.
2. If insufficient cylinder capacity exists, do not try to apply heat directly to the cylinders, and do not heat the chlorine storage room with a space heater unless the control equipment (chlorinator) room can be brought to the same temperature.
3. The chlorine cylinder should always be kept cooler than the control equipment if possible; otherwise reliquefaction of chlorine may occur at the chlorinator.

Problem – INSUFFICIENT CHLORINE GAS PRESSURE AT THE CHLORINATOR

Indicators

1. Chlorine pressure gage at chlorinator is reading too low.
2. Chlorine supply lines from cylinders are either very cold or are icing.
3. There is icing or considerable cooling at one point in the chlorine header system between the cylinders and chlorinator.

Monitoring, Analysis and/or Inspection

1. Reduce feed rate on chlorinator to about one-tenth the rotameter capacity.
2. If icing condition or cooling effect does not disappear, mark the point where cooling begins and secure the chlorine supply system at the cylinders, but let the chlorinator continue to operate.

Corrective Measures

1. When chlorine gas pressure at chlorinator reaches zero and with chlorinator still operating, disconnect flexible connection to one chlorine cylinder. (This will allow chlorinator to evacuate residual chlorine in header system by replacing chlorine with air.)
2. Disassemble chlorine header system at point where cooling began. A stoppage or a flow restriction will be found at or near this point.
3. After the stoppage has been found it can be cleaned with a solvent such as tri-chlorethylene.
4. For massive build-up in black steel pipe header systems, the pickling process

should be used. This consists of isolating the header system by disconnecting it from the cylinders at one end, the chlorinators at the other, and flushing with cold water until the water coming out is clear. The header then has to be dried with steam or hot air and final air drying to a dew point of $-40°F$.

Problem — THERE IS NO CHLORINE GAS PRESSURE AT THE CHLORINATOR, WHEN APPARENTLY FULL CHLORINE CYLINDERS ARE CONNECTED TO THE CHLORINE SUPPLY SYSTEM.

Indicators

1. Chlorinator gas pressure gage is at zero, inlet valve is open, all valves beginning with chlorine cylinder valve to the chlorinator are open.

Monitoring, Analysis and/or Inspection

1. Check the external chlorine pressure reducing valve installed just downstream of the chlorine cylinders.

Corrective Measures

1. If normal chlorine pressure appears at the chlorinator secure all the main chlorine cylinder valves, and start ventilating fans if available and arrange for maximum ventilation.
2. Put on a gas mask and gingerly break one flexible connection joint to release the gas in the header system.
3. Place a bottle of ammonia on the floor near the connection to be broken and when a white vapor appears leave the area as fast as possible and return only when the vapor disappears.
4. Repair the reducing valve whihc is probably plugged from the inherent impurities in chlorine gas. These units should be put on bi-annual overhaul.
5. Install a chlorine gas pressure gage upstream of the pressure reducing valve.

Problem —IMPOSSIBLE TO OPERATE CHLORINATOR BECAUSE ROTAMETER TUBE ICES OVER AND FEED RATE INDICATOR IS EXTREMELY ERRATIC. CHLORINE SUPPLY IS FROM TON CONTAINERS CONNECTED TO THE GAS PHASE.

Indicators

1. There is sufficient chlorine gas pressure and injector vacuum and the chlorine is at room temperature but the rotameter tube that indicates chlorine feed rate is nearly completely iced over.

2. The entire chlorine supply line back to the cylinder is also iced over, but cylinders are at about ambient temperature.

Monitoring, Analysis and/or Inspection

1. Inspect the chlorine cylinder area to see if they are connected properly. (This problem is specific to ton containers.)

Corrective Measures

1. Shut off main outlet valve on all cylinders and evacuate chlorine in header system until gage pressure at chlorinator reads zero.
2. Disconnect the cylinder that had icing on the flexible connection to the outlet valve and rotate it 180° and reconnect to top outlet valve and with other cylinders closed place this one in operation.
3. Tag cylinder so that packager can identify it as defective with possible broken dip tube but allow cylinder to remain in use until empty.

Problem – CHLORINATOR WILL NOT FEED ANY CHLORINE EVEN THOUGH ALL SYSTEMS APPEAR NORMAL.

Indicators

1. Chlorinator feed rate indicator shows little or no indication of chlorine flow when chlorine control valve is moved from closed to wide open position.
2. The chlorine pressure gage in the chlorinator is normal but the injector vacuum gage shows an abnormally high vacuum.

Monitoring, Analysis and/or Inspection

1. Check for an obstruction in the chlorine gas line near or at the inlet cartridge of the chlorine pressure reducing valve inside the chlorinator by shutting off chlorine supply system at chlorinator; chlorine pressure gage remains the same or moves downward in pressure at a very slow rate, i.e., one division per five minutes.

Corrective Measures

1. Shut off chlorine supply at the cylinders and try to let the chlorinator drain off all the chlorine gas pressure in the chlorine supply line.
2. If this cannot be done, turn on the ventilating equipment in the chlorine container space, if any, open all windows, don a gas mask and break a connection in the chlorine supply header but be absolutely sure that all chlorine cylinders have been secured.
3. When the gas has sufficiently cleared itself from the working area, disassemble

the chlorinator chlorine pressure reducing valve to remove inlet cartridge and clean stem and seat with a soft cloth.

4. If this situation occurs regularly during hot weather, the source of the trouble usually is a result of the chlorine cylinders being hotter than the chlorine control apparatus.
5. Inspect cylinder area to see if anything can be done to make the area cooler.
6. Do not connect a new cylinder if it has been allowed to sit in the sun.
7. Install an external chlorine pressure reducing valve adjacent to the last chlorine cylinder connected to the supply system.
8. Precede the reducing valve by a combination chlorine filter and sediment trap.

Problem — CHLORINE GAS IS LEAKING FROM VENT LINE CONNECTED TO EXTERNAL CHLORINE PRESSURE REDUCING VALVE (CPRV).

Indicators

1. There is no visible indication of a malfunction.
2. Chlorine escaping from CPRV vent line.
3. Chlorine gas pressure, chlorine feed rate and injector vacuum are all normal.

Monitoring, Analysis and/or Inspection

1. Confirm leak by placing ammonia bottle near the termination of the CPRV vent line.

Corrective Measures

1. The symptom described indicates that the main diaphragm of the CPRV has been ruptured.
2. Remove the external CPRV after evacuating header system and replace with a jumper tube for temporary operation while valve is being repaired.
3. Disassemble valve and replace diaphragm.
4. Inspect the ruptured diaphragm to see if failure is from corrosion, improper assembly or just fatigue from length of service.
5. Consult manufacturer for expert opinion.
6. If failure is from corrosion the chlorine supply system should be inspected for moisture intrusion.

Problem — INABILITY TO MAINTAIN CHLORINE FEED RATE
WITHOUT ICING OF CHLORINE SUPPLY SYSTEM BETWEEN EXTERNAL
CHLORINE PRESSURE REDUCING VALVE AND CHLORINATOR.
(EQUIPMENT CONSISTS OF EVAPORATOR, EXTERNAL CPRV AND
THE CHLORINATOR.)

Indicators

1. Noticeable cooling of gas line to chlorinator beginning at outlet of external chlorine pressure reducing valve.
2. Evaporator water bath temperature is normal: 160 to $180°F$.
3. Further cooling at point of pressure reduction in chlorine pressure reducing valve in chlorinator assembly.
4. Deposit of "gunk" on chlorinator feed rate indicator tube and what appears to be droplets of an amber color liquid.

Monitoring, Analysis and/or Inspection

1. Reduce feed rate on chlorinator to about 75 percent of evaporator capacity. If this or further reduction of feed rate eliminates the above symptoms, the difficulty is most likely to be insufficient evaporator capacity.
2. If chlorine gas temperature is available, calculate the superheat. If there is less than $5°F$ of superheat, there is very little reserve capacity in the evaporator. This is the result of an accumulation of sludge in the bottom of the liquid chlorine vessel of the evaporator.

Corrective Measures

1. Check for stoppage in the external CPRV cartridge if superheat cannot be measured.
2. Take the evaporator out of the system, flush and clean it with cold water and dry it in accordance with the manufacturer's instructions utilizing an "Evaporator Cleaning Kit." All evaporators should be routinely cleaned after passage of 250 tons of liquid chlorine.

Problem — CHLORINATION FACILITY CONSISTING OF EVAPORATOR-
CHLORINATOR COMBINATION WITH EXTERNAL CHLORINE PRESSURE
REDUCING AND SHUT-OFF VALVE IS UNABLE TO MAINTAIN WATER-
BATH TEMPERATURE SUFFICIENT TO KEEP EXTERNAL CHLORINE
PRESSURE REDUCING VALVE IN OPEN POSITION

Indicators

1. External chlorine pressure reducing valve shuts off intermittently until water
   bath temperature is raised about $150°F$.
2. Intermittent operation of chlorination equipment
3. Insufficient heat being supplied to evaporator water bath.

Monitoring, Analysis and/or Inspection

1. Check evaporator water bath temperature.

Corrective Measures

1. After evaporator has been in operation sufficiently long enough to bring heating
   elements to operating temperature, shut down power supply and remove and
   replace heating elements.

Problem — INABILITY TO OBTAIN MAXIMUM FEED RATE FROM
CHLORINATOR OR CHLORINATORS WITH ADEQUATE CHLORINE GAS
PRESSURE AT CHLORINATOR

Indicators

1. Chlorinator is placed into manual control and control valve is opened wide but
   chlroine feed rate will not go beyond 70 to 80 percent of maximum.
2. Check injector vacuum gage to see if reading is less than minimum recommended
   by manufacturer.
3. Check for injector vacuum reading below ten inches Hg.

Monitoring, Analysis and/or Inspection

1. Reduce feed rate on chlorinator.
2. If the injector vacuum reading increases then increase the injector water
   pressure.
3. Verify whether or not inlet water pressure to the injector is the same as when
   the installation was first installed.

4. Check injector water pump pressure against the manufacturer's operating data.

Corrective Measures

1. Disassemble injector and see that the throat and tailway are clear and without any abnormal deposition of iron or manganese, and clean the injector parts by soaking in muriatic acid, rinse in fresh water and replace.

Problem — INABILITY TO MAINTAIN ADEQUATE CHLORINE FEED RATE

Indicators

1. Inspection of chlorinator reveals that chlorinator cannot feed as much as previoulsy noted even though chlorine supply pressure is adequate.

Monitoring, Analysis and/or Inspection

1. If effluent is used check the injector operating water supply for deterioration in supply pump performance.

Corrective Measures

1. In the case of a centrifugal pump the only solution is a complete overhaul.
2. If a turbine pump, close down on the needle valve to maintain the proper discharge pressure.
3. If the turbine pump has worn sufficiently and it requires operation with the needle valve in the fully closed position, the pump should be thoroughly overhauled.

Problem — INABILITY TO OBTAIN MAXIMUM OR PROPER FEED RATE FROM CHLORINATOR WITH ADEQUATE GAS PRESSURE AT CHLORINATOR

Indicators

1. With chlorinator in manual control and chlorine control valve is manipulated to very the feed rate, the change of feed rate response seems sluggish and chlorinator will not achieve maximum feed rate.
2. The injector vacuum reading is borderline, and when feed rate is reduced the injector vacuum does not increase appreciably.

Monitoring, Analysis and/or Inspection

1. Check the chlorinator vent system for a small vacuum leak in the chlorine control apparatus by disconnecting the vent line at the chlorinator and while

observing the chlorinator operation (feed rate and injector vacuum), place a hand over the vent connection to the vacuum relief device on the chlorinator. If this action produces more injector vacuum and more chlorine feed rate, it signifies that air is entering the chlorinator via this mechanism (vacuum relief device) because the springs have become weak due to normal metal fatigue.
2. Moisten all joints subject to a vacuum with ammonia solution or put paper impregnated with orthotolidine at each of these joints. With chlorinator operating at maximum feed rate, close the injector discharge line as rapidly as possible. If there is a vacuum leak in the chlorinator system it will be detected by either the ammonia or the paper.

Corrective Measures

1. If the vacuum leak is in the vacuum relief device, disassemble the mechanism and replace all the springs.
2. Repair all other vacuum leaks by tightening a joint, replacing gaskets, replace tubing and/or compression nuts.

Problem — EXCESSIVE CHLORINE ODOR AT POINT OF APPLICATION

Indicators

1. Air cover above area of chlorine diffuser reacts with ammonia solution to produce typical white wisps of "smoke" indicating escaping molecular chlorine.

Monitoring, Analysis and/or Inspection

1. Scattering ammonia indicator solution onto the wastewater stream over the area of the diffuser produces white fumes at the surface.
2. Check chlorine solution strength.

Corrective Measures

1. Add enough injector water to bring the chlorine solution strength down to 3500 ppm chlorine at maximum expected chlorine feed rate.
2. If the chlorine diffuser is situated below the injector which leads to a negative head in the solution line, install a special diaphragm protected chlorine solution pressure gage in the highest point of the chlorine solution discharge line and regulate the injector water flow so that there is a 2 to 3 psi positive pressure at this point.

Problem — CHLORINATOR WILL NOT FEED ENOUGH CHLORINE TO
PRODUCE A PROPER CHLORINE RESIDUAL AT THE SAMPLING POINT.

Indicators

1. A routine spot check sampling shows that at some hours of the day there is an adequate residual but there are times during the day when there is no residual.
2. If there is a chlorine residual analyzer the chart will show periods during the day of insufficient chlorine residual.

Monitoring, Analysis and/or Inspection

1. Ascertain that if the chlorination equipment is being used for disinfection that it is equipped to proportion the chlorine feed rate in accordance with the flow of the wastewater.
2. If it is flow proportional, check to see if the meter capacity on the chlorinator matches the plant flow meter capacity.
3. Disconnect the flow proportional control and by manual control test the chlorinator to see if it will pull maximum feet rate.
4. Determine if solids have settled to the bottom of the contact chamber.

Corrective Measures

1. The automatic control features of the chlorinator should be repaired by the manufactuerer's field service personnel who are equipped to simulate the various types of electric and pneumatic signals commonly used for chlorinator control.
2. If needed, clean the chlorine contact chamber.

Problem — WIDE VARIATION IN CHLORINE RESIDUAL IN EFFLUENT
AS DETERMINED BY HOURLY CHLORINE RESIDUAL DETERMINATIONS

Indicators

1. Inability to adjust dosage so that there is reasonable agreement of chlorine residual throughout a 24-hour period as determined by occasional chlorine residual analysis at each shift.

Monitoring, Analysis and/or Inspection

1. While in flow proportional operation the feed rate of the chlorinator should be plotted on a piece of graph paper against the flow meter reading. The plots of wastewater flow versus chlorine feed rate should yield a straight line.

Corrective Measures

1. If the feed rate plots do not follow a reasonably straight line, it is well to

recheck the zero and span of the flow proportional control device on the chlorinator. First make the zero check and then the span check in accordance with the manufacturer's instructions. If this does not correct the difficulty, it may be necessary to replace operating parts within the controller to achieve satisfaction.
2. If after the flow proportional control system on the chlorinator has been corrected the irregular chlorine residual reading continues, then it is recommended that a continuous chlorine residual analyzer be installed.

Problem — CHLORINE RESIDUAL ANALYZER RECORDER CONTROLLER DOES NOT APPEAR TO CONTROL THE CHLORINE RESIDUAL PROPERLY.

Indicators

1. Recorder draws a poor line on the chart that seems not to bear any relation to the "set point."

Monitoring, Analysis and/or Inspection

1. First check the loop-time in the system. This is best accomplished by turning off the gas supply to the chlorinator and determining the length of time required to show a sharp drop in the residual on the analyzer chart.
2. Disconnect the analyzer cell output leads from the cell and apply a simulated signal to the recorder mechanism from a manually controlled external signal generator. (Authorized chlorinator repair personnel carry such a device as part of their tool kits.)
3. Check buffer additive system to see if pH of sample going through the cell is maintained at 5 or less.
4. Check electrode bombardment system and see that electrodes are clean, particularly the noble metal electrode (Pt. or Au.). Do not distrub the copper electrode unless it is fouled with grease.
5. If residual analyzer is being used to measure total residual, check to see if sufficient potassium iodide is being added for the amount of residual being measured.
6. Reconnect cell output leads and make a zero, span and temperature check by following the manufacturer's procedure for a routine calibration.

Corrective Measures

1. If the loop time is found to be in excess of 5 minutes, satisfactory operation will not be achieved until the loop time is brought down to 5 minutes or less. This can be accomplished by moving the injector closer to the point of application, increasing the velocity in the sample line to the analyzer cell, by moving the cell closer to the sample point, or by moving the sample point closer to the point of application.

2. If the line on the chart indicates proper operation when subjected to a simulated signal, this signifies that the equipment between the cell and the readout of the pen is satisfactory. The erratic or poor line can either be caused by poor mixing of chlorine at the point of application or faulty operation of the cell. Poor mixing can be verified by setting the chlorine feed rate for a constant dosage (proportional to flow) and analyzing a great many grab samples over a ten minute period as quickly as possible. A poor mix will show rapid wide swings of the recorder pen. Consider mixing the point of application and/or install some type of mixing device to cause turbulence at the point of application.

3. If poor mixing is not the cause, and if the electrodes are clean, and if the pH and KI additive system is normal, then the difficulty must be in the cell and it should be replaced.

4. If when the simulated signal is applied to the recorder mechanism and the recording system does not respond properly, the difficulty lies in the electrical components of the recorder mechanism. Authorized service personnel should be summoned to correct the difficulty.

Problem — CHLORINATION SYSTEM CONSISTS OF EITHER COMPOUND-LOOP CONTROL OR DIRECT RESIDUAL CONTROL AND SYSTEM DOES NOT APPEAR TO BE CONTROLLING PROPERLY

Indicators

1. Chlorine residual line on analyzer chart appears normal but does not track close enough to set point.

Monitoring, Analysis and/or Inspection

1. If the chlorine residual analyzer is operating properly, check the chlorinator system to see that it is functioning properly over its entire range of feed rate. Check to see if the chlorination system is feeding enough chlorine to satisfy the maximum demand; also check to see if the rotameter tube range is sized so that incremental corrections in feed rate by the residual controller are not too large. These two factors would cause wide swings in the chart line, or not allow the chlorine applied to ever actually "catch up" with the set point.

Corrective Measures

1. If the chlorination system will not feed enough chlorine, consult the corrective measures described previoulsy under the problem of chlorination control equipment unable to feed enough chlorine.

2. If the chlorinator rotameter tube range gives too large or too small an incremental change, replace with a proper range of feed rate.

Problem — COLIFORM COUNT DOES NOT MEET THE REQUIRED
DISINFECTION STANDARDS SET BY REGULATORY AGENCIES.

Indicators

1. Routine analysis of effluent or receiving waters shows MPN coliform organism
   to be in excess of that required by regulatory authorities.

Monitoring, Analysis and/or Inspection

1. Check capacity of chlorination equipment as follows: For primary effluent
   chlorinator capacity should be from 175 to 200 lb per MG. For secondary
   effluent 100 to 125 lb per MG and for tertiary effluent 75 to 100 lb per MG
   unless nitrogen removal is required. For the latter or for those plants requiring
   free residual chlorine, equipment capacity must be 10 mg/1 of chlorine for each
   mg/1 ammonia nitrogen in the effluent.
2. All chlorination equipment used for disinfection of wastewater effluent should
   have at the very least control proportional to the effluent flow. The capacity of
   the chlorinator should also be based on the maximum reading of the flow meter.
3. Continuously record the residual in the effluent with an amperometric type
   chlorine residual analyzer.
4. Check for short circuiting in contact chamber.

Corrective Measures

1. Chlorination equipment should be brought up to optimum capacity requirements. The necessary equipment should be installed to provide flow proportional control. In plants where only an influent meter exists, it may be required
   to install an effluent meter. After the proper primary meter is installed then the
   chlorinator can be modified by adding a chlorine orifice positioner to be operated either electrically or pneumatically from the primary meter.
2. A chlorine residual analyzer should be installed to properly monitor the chlorine
   control system. Using this apparatus to automatically control the chlorine
   dosage is optional; however, experience shows that the change in chlorine demand of most domestic wastewaters is significant enough to warrant the small
   added expense to accomplish automatic dosage control.
3. Install additional baffling in contact chamber.
4. If needed, install a mixing device in contact chamber.

Problem — COLIFORM COUNT DOES NOT MEET THE
REQUIRED STANDARDS FOR DISINFECTION.

Indicators

1. Routine analysis of effluent or receiving waters shows MPN of coliform organ-

isms to be in excess of the required standards.

Monitoring, Analysis and/or Inspection

1. Check to see if chlorine capacity is adequate, control system is functioning properly and effluent is being monitored with a continuous chlorine residual analyzer.
2. Check the chlorine contact time at low flow, average flow and maximum flow to determine the optimum residence time of the process. With this as a basis analyze five replicate samples for each hour around the clock on Monday, Wednesday, Friday and Sunday for coliform MPN after the following treatment: samples are to be taken from the effluent prior to point of application of chlorine and dosed in the laboratory with the same amount of chlorine as that applied by the chlorination equipment. The chlorine for this procedure should be taken from the plant chlorine solution line, standardized according to Standard Methods and added to one liter replicate sample of effluent. Upon addition the chlorine solution should be rapidly and thoroughly mixed, then allowed to stand for the amount of time determined previously at the optimum residence time. At the expiration of the residence time one portion of the samples should then be analyzed for chlorine residual using the iodemetric back titration procedure while another portion should be dechlorinated and analyzed for coliform MPN in accordance with Standard Methods.
3. If the resulting coliform MPN from the above analysis is satisfactory, it is then reasonable to assume that the mixing at the point of application is at fault, because stirring chlorine in a batch process described above results in ideal chemical mixing.
4. Check for solids buildup in contact chamber.

Corrective Measures

1. If the difficulty is too low a residual raise feed rate and increase contact time if possible.
2. If poor mixing is the problem install a mixing device of high turbulence such as exists in an hydraulic jump or a combination of turbulent flow and mechanical mixer.
3. Clean contact chamber to reduce solids buildup.

Problem —PLANT EFFLUENT DOES NOT MEET TOXICITY REQUIREMENTS BECAUSE CHLORINE RESIDUAL TO ACHIEVE PROPER DISINFECTION IS AT TOO HIGH A LEVEL.

Indicators

1. Toxicity level is too high as determined by present bio-assay procedures.

Monitoring, Analysis and/or Inspection

1. Chlorine residual as determined by iodometric method using back titration method is deemed toxic to fish and other aquatic life in the receiving waters.

Corrective Measures

1. Install a dechlorination facility to operate in conjection with the chlorination system.

VI. METERING

Problem — PLANT METER UNREALIABLE

Indicators

1. Drop or sharp increase in totalized dry weather flow
2. Overly uniform flow chart

Monitoring, Analysis and/or Inspection

1. If meter operates on a float, check float well for obstructions.
2. If meter operates on bubbler, check bubbler tube for damage. Also check air pressure gage to see that meter is getting proper air flow.
3. Bypass measuring weir on flume if possible and check to see if meter zeros.
4. Check height of flow over weir or in flume at different time intervals, and, using the weir or flume characteristics formulas, calculate flow and compare with flow meter data.
5. Compare water surface elevation immediately behind weir or flume with elevation of water in float or bubbler well.
6. Ascertain the fact that none of the plant's process return flows (centrate, supernant, waste activated sludge, etc.) are discharged upstream from the meter.
7. Install a portable flow meter in the weir or flume and compare results with plant flow.
8. If metered flow discharges into a wet well or other chamber which has a known volume and outflow can be shut off, record time of measured rise in chamber and calculate inflow rate. Compare calculated data with meter data.
9. If wastewater treated at the plant is supplied by one or more utilities and the total amount of water used is metered, compare area water consumption with plant flow. During fall or early winter months, water consumed is approximately 10% more than wastewater discharged.
10. Check magnetic flow meter cores for grease build-up or restrictions.

Corrective Measures

1. Keep all floats and bubbler wells clean and free of grease by periodic maintenance.
2. Differences between inside and outside or bubbler well water surface elevations are due to extreme velocities in the immediate area of these wells. If possible, move wells to a more quiescent area behind the weir or flume.
3. Clean all foreign matter off weir plates.
4. If problem appears to be in the meter, recording or telemetering equipment, a qualified technician should be called in to repair and calibrate the metering equipment.

VII. SOLIDS HANDLING — Sludge Thickeners

Problem — ODOR FROM THICKENER

Indicators

1. Odors of hydrogen sulfide origin
2. Floating or gaseous sludge in thicknener
3. Corrosion of thickener concrete structure and metal work

Monitoring, Analysis and/or Inspection

1. Run total and dissolved sulfide test of thickener effluent.
2. Check pumping rate and frequency of pumping raw sludge from thickener.
3. Run total solids test on raw sludge pumped.
4. Dewater thickener and check operation of a scrapper and/or stirrer arms and sludge removal equipment.
5. Determine sludge blanket depth.

Corrective Measures

1. Adjust pumping rate to remove solids at a frequent rate and at not less than 3% total solids.
2. Repair or replace all damaged sludge collector mechanisms.
3. Cover thickener and exhaust gases to an odor control scrubber.

Problem — THICKNENER CONTENTS DO NOT SETTLE

Indicators

1. Floating sludge on thickener surface
2. Floc in thickener effluent

3. Increased loadings on primary sedimentation tanks and secondary treatment process
4. Excessive sludge solids in the overflow
5. Poor concentration of underflow.

Monitoring, Analysis and/or Inspection

1. Run total solids on thickener effluent.
2. Run total solids in raw sludge withdrawn from thickener.
3. Run total solids of all thickener inflows and compare to thickener design capacity.
4. Run 30 minute settleability test of waste activated sludge inflow to thickener.

Corrective Measures

1. If thickener effluent contains high solids and the raw sludge withdrawn low solids, dose thickener with polymers or other coagulants.
2. If solids loading exceeds thickener design capacity, partially bypass thickener, if possible, by pumping raw sludge from the primary sedimentation tanks directly to the point of disposal, digester, or incinerator.
3. If waste activated sludge pumped to thickener does not readily settle, re-aerate or treat with coagulants.

Problem — SLUDGE PUMPED FROM THICKENER HAS
LOW SOLIDS CONCENTRATION

Indicators

1. Thin or watery sludge discharged to point of disposal

Monitoring, Analysis and/or Inspection

1. Run total solids of raw sludge pumped from thickener.
2. Check pumping cycle and rate of pumping raw sludge from thickener.
3. Check rate of inflow to thickener.
4. Run total solids of thickener inflow.
5. Determine depth of the sludge blanket.

Corrective Measures

1. Adjust pumping rates and cycles, perferably with timers, and remove raw sludge from thickener at a density not less than 3% total solids.
2. Adjust thickener inflows to apply total solids to thickener of not less than 2%.
3. In gravity thickness adjust pumping cycles to maintain 3 to 4 ft sludge blanket.

Problem — SCUM BLANKET IN TANK

Indicators

1. Decrease in digester gas production
2. Crust visible through sight glasses in digester roof
3. Unable to supernate from upper level of digester

Monitoring, Analysis and/or Inspection

1. Core blanket through digester thief holes to determine thickness.
2. Check digester temperature.
3. Check daily digester gas production.
4. Determine gallons of scum pumped to digester daily.

Corrective Measures

1. If possible, recirculate digested sludge from bottom of digester to top of scum blanket.
2. If digester has a gas mixing system, run system continuously while increasing digester temperature to not more than 105°F with the incremental increases not exceeding 1°F per day.
3. If digester has mechanical mixers with draft tubes, degassify digester, break up scum with a high pressure water jet and direct to draft tubes.
4. Clean digester and find alternate means of scum disposal.
5. If digester has gas mixer system, place temporary gas diffusers in thief holes and pipe compressed digester gas to them.

Problem — NO DIGESTER GAS PRODUCTION

Indicators

1. Gas produced has septic odor.
2. Gas produced does not ignite
3. Increase in digester volatile acids
4. Increase in volatile acid/alkalinity ratio

Monitoring, Analysis and/or Inspection

1. Determing digester volatile acid, alkalinity, and pH of digested sludge together with trend of volatile acid/alkalinity ratio
2. Check gas meter and piping for restrictions.
3. Monitor volume of raw sludge pumped to digester daily.

4. Determine total solids, volatile solids, and pH of raw sludge pumped to digester.
5. Check digester temperature.
6. Sound digester to determine depth of scum blanket and grit residue on bottom.
7. Calculate volatile matter reduction in digester.
8. Check for toxic material in the digester.

Corrective Measures

1. If volatile acid to alkalinity ratio is greater than 0.2 and pH below 6.5 add lime to digester to decrease volatile acid/alkalinity ratio and increase pH.
2. Do not feed digester raw sludge in low pH ranges (less than 7.0).
3. If volatile reduction in digester is less than 50% decrease or discontinue feeding digester until pH rises.
4. Do not feed digester raw sludge with average volatile solids less than 75%.
5. If possible transfer digested sludge with a volatile acid/alkalinity ratio of 0.2 from another digester to affected digester.
6. If scum blanket and/or grit deposits comprise more than 50% of the effective volume of the digester, clean the tank.
7. Keep digester temperature at 98°F.
8. Clean all restrictions in gas lines and/or meters.
9. If toxic material has killed digester, clean digester and determine source of toxicity to prevent recurrence.

Problem — INCREASE IN VOLATILE ACID/ALKALINITY
RATIO IN DIGESTER

Indicators

1. Drop in digester gas production
2. Hydrogen sulfide odor from digester supernatant.

Monitoring, Analysis and/or Inspection

1. Determine volatile acid, alkalinity and pH of digested sludge at least twice daily.
2. Check digester temperature.
3. Check pH of raw sludge pumped to digester.
4. Check mixing in digester.

Corrective Measures

1. If digester pH is below 6.5, add lime to digester.
2. If volatile acid/alkalinity ratio is greater than 0.4 decrease or discontinue feeding digester and add lime.
3. Do not feed digester raw sludges with pHs lower than 6.8.

4. Do not let digester temperature drop below 90°F.
5. If possible, transfer sludge with low volatile acid/alkalinity ratio content from another digester to affected digester.
6. Keep contents of digester well mixed.
7. Decrease sludge withdrawal rates from digester.

Problem — FOAM IN DIGESTER

Indicators

1. Foam discharged from upper level supernatant lines
2. Froth visible through sight glasses in digester roof

Monitoring, Analysis and/or Inspection

1. Determine total and volatile solids of sludge being pumped to digester and volume pumped.
2. Determine pH of digester contents.
3. Check digester temperature daily.
4. Monitor withdrawal rate of sludge from digester.
5. Acertain depth and/or thicknesses of grit deposits and/or scum layers.
6. Check digester mixing program and effectiveness of mixing equipment.

Corrective Measures

1. Maintain digester pH between 6.8 and 7.2 and volatile acid/alkalinity ratio below 0.2 by adding lime.
2. Reduce or discontinue pumping raw sludge to digester.
3. Maintain digester temperature constant and at least at 95°F.
4. Attempt to thoroughly mix digester by recirculation or by available digester mixing equipment.
5. Break up scum layers or, if not possible, clean digester.
6. If possible, add digested sludge from a healthy digester.

Problem — LOW REDUCTION OF VOLATILE SOLIDS IN DIGESTER

Indicators

1. Volatile reduction calculates to less than 50%

Monitoring, Analysis and/or Inspection

1. Determine total solids of digested sludge and/or raw sludge being pumped to digester.

2. Monitor solids loading to digester daily.
3. Monitor solids withdrawal from digester.
4. Check total solids in digester supernatant.
5. Acertain depths and/or thickness of grit deposits and/or scum layers.
6. Determine volatile acid/alkalinity ratio and pH of digested sludge.
7. Monitor digester gas production.

Corrective Measures

1. If total or volatile solids daily loading of digester exceeds design loading, reduce the amount of sludge pumped to the digester daily.
2. Keep digester temperature about 95°F.
3. Raw sludge pumped to digester should contain more than 50% volatile solids.
4. Recirculate and mix digester
5. Prolong periods of withdrawing digested sludge until volatile reduction is above 50%.
6. Lower volatile acid/alkalinity ratio and raise pH above 6.5 by adding lime to digester.
7. If supernatant contains high solids content, let digester settle.

Problem — HIGH PERCENT SOLIDS IN DIGESTER SUPERNATANT

Indicators

1. Supernatant very dark and thick
2. If supernatant is circulated to plant headworks, primary and/or secondary treatment processes efficiency drops severely.

Monitoring, Analysis and/or Inspection

1. Determine total solids of digester supernatant while supernating from different levels in the digester.
2. Determine total solids of digested sludge.
3. Monitor amount of raw sludge, activated sludge, and scum pumped to the digester daily.
4. Check pH of digester supernatant.
5. Determine length of time of digester mixing.

Corrective Measures

1. If supernatnat contains more than 1% solids, do not mix digester and feed alternate digester if possible.
2. Supernate from digester level which gives the least solids in the supernatant.
3. Supernate from one digester into another if possible.

Problem — LOW SOLIDS RECOVERING RATE

Indicators

1. Centrifuge efficiency falls below 60%
2. Solids in centrate exceed 3%
3. Centrate very dark in color
4. If centrate is recirculated to the plant headworks, the efficiency of primary and secondary treatment processes are directly affected.
5. Cake appears thick and quite wet.

Monitoring, Analysis and/or Inspection

1. Calculate centrifuge efficiency.
2. Monitor sludge feed rate to centrifuge and percent solids in sludge.
3. Monitor coagulant, if any, feed rate to centrifuge.
4. Check centrifuge for mechanical wear.
5. Check pool depth.

Corrective Measures

1. Decrease sludge feed rate to centrifuge.
2. Increase chemical coagulant dosage, if any.
3. Repair or replace all worn centrifuge parts.
4. Increase pool volume and bowl speed.
5. Reduce conveyor speed.

VII. SOLIDS HANDLING — Vacuum Filters

Problem — LOW SOLIDS RECOVERY

Indicator

1. High solids in filtrate
2. Poor clarity filtrate

Monitoring, Analysis and/or Inspection

1. Calculate filter efficiency.
2. Monitor coagulant, if any, feed.
3. Check filter mesh for blindings or coarseness.

Corrective Measures

1. Increase chemical coagulant feed rate.
2. Clean filter media.
3. Install fine mesh filter media.
4. Check sludge washing (elutriation) process.
5. Check filter drum speed and operation cycle.
6. Change types of coagulants being used.

VII. SOLIDS HANDLING — Incineration

Problem — ABNORMALLY HIGH TEMPERATURE IN FURNACE

Indicators

1. Temperature indicator exceeds limit of maximum operating temperature.
2. High temperature alarm activated.

Monitoring, Analysis and/or Inspection

1. Check rate of fuel consumption to determine if excessive.
2. Check to determine if fuel feed is off and temperature is still rising. Greasy solids may be present.
3. Check temperature indicator to see if it reads all the way up on the scale.

Corrective Measures

1. Decrease fuel feed rate, if excessive, in relation to sludge feed rate.
2. If temperature rises without supplementary fuel feed, greasy solids may be present in sludge. (This occurs infrequently when 100% primary sludge is incinerated.) To lower temperature, raise air feed rate while holding sludge feed rate constant. If air feed rate is at maximum, reduce sludge feed rate slightly.
3. If temperature indicator is all the way up the scale, this is an indication that the thermocouple well is burned out. Replace thermocouple with spare unit, repair or replace thermowell with spare.

Problem — ABNORMALLY LOW TEMPERATURE IN FURNACE

Indicators

1. Temperature indicator shows low on scale.

Monitoring, Analysis and/or Inspection

1. Check calorific value of sludge to determine if it is decreasing.

2. Check moisture content of sludge to determine if it is increasing.
3. Determine level of excess oxygen in stack.

Corrective Measures

1. If calorific value of sludge is low, or if sludge moisture content is high, increase the supplementary fuel feed rate.
2. If excess oxygen is abnormally high in stack exhaust, reduce the air feed rate slighty or increase sludge feed rate.

Problem — HIGH OXYGEN LEVEL IN FURNACE STACK EXHAUST

Indicators

1. Oxygen analyzer (recorder chart) indicates excessive oxygen in stack exhaust

Monitoring, Analysis and/or Inspection

1. Determine total and volatile solids of sludge fed to furnace.
2. Check to determine if sludge is being fed to incinerator.

Corrective Measures

1. If sludge being fed to furnace is low in solids, increase the speed of sludge pump which feeds centrifuge. If the air feed rate is low and the sludge rate at a maximum and the exhaust oxygen is still high, shut down. The sludge supply to the furnace is inadequate.
2. If sludge is not being fed to incinerator, check for blockage of sludge in feed chute and check sludge feed pump stator. Unplug chute or repair stator if not in good condition.

Problem — LOW OXYGEN LEVEL IN FURNACE STACK EXHAUST

Indicators

1. Oxygen analyzer (recorder chart) indicates low oxygen in stack exhaust.

Monitoring, Analysis and/or Inspection

1. Check volatile content of sludge fed to furnace to determine if increasing.
2. Check sludge for grease content to determine if increasing.
3. Check to determine if air flow to furnace is restricted.

Corrective Measures

1. If the volatile content of the sludge is increasing or if the grease content of the

sludge is increasing, increase the air feed rate. If the air feed rate is at maximum decrease sludge feed rate.
2. Remove any restrictions or blockages in air conduits.

VII. SOLIDS HANDLING — Sludge Lagooning

Problem — EXCESSIVE SOLIDS CARRIED OVER FROM LAGOON SUPERNATANT TO PLANT INFLUENT

Indicators

1. Dark supernatant discharged from sludge lagoons
2. Solids removal efficiency of plant treatment processes is lowered.

Monitoring, Analysis and/or Inspection

1. Determine total and suspended solids of lagoon supernatant.
2. Measure depth of sludge in lagoons.
3. Check for broken dikes between lagoons.
4. Check rate and volume of application of digested sludge to lagoons.
5. Determine total solids of sludge being applied to lagoons.

Corrective Measures

1. Reduce volume of sludge applied to lagoons thereby reducing depth of sludge in lagoon.
2. Repair all broken dikes between lagoons.
3. Delay release from lagoons of supernatant with heavy solids content until sludge is allowed to settle.

Problem — ODORS FROM SLUDGE LAGOONS

Indicators

1. Obnoxious odors from sludge lagoons.

Monitoring, Analysis and/or Inspection

1. Determine volatile acid/alkalinity ratio and pH of digested sludge being applied to lagoons.
2. Determine total and dissolved sulfide of lagoon supernatant.

Corrective Measures

1. If volatile acid/alkalinity and pH tests indicate a sour digester, attempt to correct

problem at source.

2. Apply lime to surface of lagoon.
3. Install peripheral odor control system.
4. Flood lagoon with heavy chlorinated water.

CHAPTER 7

PLANT SAFETY

INTRODUCTION

Safety in plant operations is of major importance. Unsafe procedures result in injuries, with consequent human suffering, loss of human resources, equipment breakdown, increased equipment replacement, and higher labor costs.

Any injuries that occur should be investigated in order to determine the cause and prevent the injury recurring. This fact-finding procedure focuses attention on inadequacies of equipment and operations that exist, and also increases safety conciousness of plant personnel. An investigation must include the description of the accident, the agent responsible for the accident and/or injury, and suggest ways to avoid a recurrence. Records of investigations help establish a better safety program and point out injury-prone locations and additional preventive measures.

In general, physical injuries can be prevented by:

1. Maintaining all structures and equipment in good repair and in orderly condition.
2. Practicing good housekeeping.
3. Keeping walkways guarded and free from oil and grease.

4. Using equipment only for the use intended.
5. Taking special care near electrical and mechanical equipment.

Prevention of Body Infections

Workers who come into contact with sewage are exposed to the hazards of waterborne diseases including: typhoid fever, paratyphoid fever, amoebic dysentary, infectious jaundice and other intestinal infections. There is also the danger of tetanus.

Prompt medical attention to any injury with accompanying skin rupture minimizes the risk of infection. A first aid kit should be available for immediate treatment of minor cuts and wounds.

Kinds of Hazards

You are equally exposed to accidents whether working on the collection system, or working in a treatment plant. As a worker, you may be exposed to:
1. Physical injuries.
2. Infections and infectious diseases.
3. Oxygen deficiency.
4. Toxic of suffocating gases or vapors.
5. Radiological hazards.
6. Explosive gas mixtures.
7. Fire.
8. Electrical Shock.
9. Noise.

The following table is indicative of the types of hazards that may be encountered in the various sections of a sewage treatment plant.

Sewers and Manholes

When it is necessary to enter a sewer manhole, the following precautions should be observed. When one person enters a manhole, someone should be available at the top to assist him in case an emergency arises. Only safety, explosion-proof, electric light equipment; or if possible, mirrors should be used to supply light. No smoking or open flames should be allowed, and personnel should guard against sparks. Rubber soled shoes or boots will help to prevent slips and infections.

Chlorination Facilities

Chlorine is generally handled in cylinders at the smaller plants and in one ton containers for systems with a high demand. If handled properly, there is very little risk involved. In the air, however, at normal temperatures and pressures, chlorine is a poisonous yellow-green gas which has a strong, pungent odor. Even relatively low concentrations in the air can cause eye irritation, coughing and labored breathing.

Table 1. Hazards Encountered at Sewage Treatment Plants.

Unit Operations At A Typical Sewage Treatment Plant	Infections/ Infec. Diseases	Electrical Shock	Noise	Oxygen Deficiency	Radiological Hazard	Fire	Explosive Gas Mixtures	Physical Injuries
Inlet Structure								
--Bar Screen	x	-	-	-	-	x	x	x
--Comminutor	-	x	x	-	-	x	x	x
Clarification Facilities	x	x	-	-	-	-	-	x
Aeration Facilities	x	x	x	-	-	-	-	x
Disinfection Facilities	-	x	-	x	-	x	x	x
Laboratory	x	x	x	-	x	x	x	x
Control Structures	-	x	x	x	-	x	x	x
Office Building	-	-	-	-	-	-	-	x
Sludge Disposal								
--Digestors	x	x	-	x	-	x	x	x
--Dewatering Equipment	x	x	x	-	-	-	-	x
--Sludge Drying Beds	x	-	-	-	-	-	-	x
Outside Facilities								
--Manholes	x	x	-	x	-	x	x	x
--Pump Stations	x	x	-	x	-	x	x	x
--Lagoons	x	-	-	-	-	-	-	x

In the presence of moisture, chlorine gas is highly corrosive. Accordingly, every precaution should be taken to avoid leaks and to repair them promptly when they occur.

The chlorinator room is a potentially hazardous area in view of the characteristics of gaseous chlorine. Before entering the chlorinator room, personnel should view the room from the outside to be sure that the yellow-green gas indicative of a chlorine leak is not present. The chlorinator room ventilation system should be turned on for at least three minutes before the room is entered, and it should remain in operation as long as anyone stays in the room.

Should it be necessary to enter the chlorinator room while chlorine gas is present, the operator must use a canister type gas mask. The canister should contain a chemical which absorbs the chlorine, and no separate oxygen supply should be required. The length of time that the mask can be worn is limited by the concentration of the contaminant and the exertion of the wearer. It is important that the manufacturer's instructions be followed carefully, and that a record be kept of mask usage to insure that the mask will be serviceable when needed. A spare canister should always be available.

For information regarding the handling and storage of chlorine containers, see the "Chlorine Manual", published by the Chlorine Institute, Inc.

Laboratory Safety

In the course of performing the routine laboratory control tests, personnel work with both sewage and chemicals. The potential hazards involved in handling sewage have been discussed previously.

Due care should be taken when handling chemicals, particularly sulfuric acid. It is recommended that a protective laboratory coat be worn when performing tests. Goggles or glasses for eye protection should be worn when handling acids or other caustic chemicals.

The operator should be particularly careful not to swallow either sewage or chemicals when pipetting. Rubber suction bulbs can be obtained for use in pipetting if the operator so desires. Potable water for rinsing eyes, mouth or body is available from the faucet in the laboratory.

Fire Hazards

Dropping lighted matches or burning tobacco, using open flames in or around sewer manholes and treatment units is dangerous and must be avoided. The presence of an explosive gas or flammable liquid, such as gasoline floating at the surface of the clarifier, can cause a serious explosion or fire.

The treatment plant and pumping station should be equipped with a number of fire extinguishers. Flammable liquids, gases and greases and electrical fires can be effectively controlled with carbon dioxide (CO_2) or dry chemical type extinguishers.

Mechanical and Electrical Equipment

The operator should become thoroughly familiar with the manufacturer's instructions regarding the proper operation and maintenance procedures for each piece of mechanical equipment in the plant. Proper operation and maintenance of a piece of machinery is the first step in assuring safety when working with or near that piece of equipment. When working on a piece of mechanical equipment, all power to the equipment should be shut off by opening the proper control switch and tagging it to prevent others from closing it. Under no circumstances should the operator attempt to perform preventive or corrective maintenance on machinery that is operating, unless otherwise indicated in the manufacturer's instruction manual.

Work should not be done on a piece of electrical equipment while standing on a wet or damp floor. A rubber mat on the floor in front of electrical panels is an added precaution. Metal ladders should never be used around electrical equipment, and all electric tools should be grounded.

Recommended Safety Equipment

The following is a list of the minimum recommended safety equipment which should be available at the treatment plant. Any other pieces of safety equipment which the operator deems necessary should be added to the list.

1. Carbon dioxide or dry chemical type fire extinguishers accessible to all hazardous locations. A fire extinguisher should also be provided for the sewage pumping station.
2. First aid kit. A first aid kit should also be provided for the sewage pumping station.
3. Gas mask with spare canister located near the entrance to the chlorine room. A gas mask and spare canister should also be provided near the chlorine facilities at the sewage pumping station.
4. "No Smoking" signs located in all hazardous locations at both the treatment plant and sewage pumping station.
5. Safety, explosion-proof lighting equipment for use when working in hazardous locations.
6. Protective clothing and goggles for use when working with chemicals.
7. Rubber mats in front of all electric panels at both the treatment plant and sewage pumping station.

Safety References

The following safety references should be obtained and kept in the Operator's office:

1. Water Pollution Control Federation Manual of Practice No. 1 "Safety in Wastewater Works"
2. Water Pollution Control Federation Manual of Practice No. 2 "Operation of Wastewater Treatment Plants"
3. Chlorine Institute "Chlorine Manual"
4. Fisher Manual of Laboratory Safety. Fisher Scientific Co. 1972.

APPENDICES

A. Glossary of Terms

B. Suggested Lubrication Materials

C. Pumps for Pollution Control and Symptoms of Pump Operating Problems

D. Valves, Pipes and Fittings

E. Preparation for a Site Visit

F. Sewer System Evaluation

G. Basic Laboratory Procedures

GLOSSARY*

Absorption (ab-SORP-shun): Taking in or reception of one substance into the body of another by molecular or chemical action, and distributed throughout the absorber.

Activated Sludge (ACK-ta-VA-ted sluj): Sludge particles produced in raw or settled wastewater (primary effluent) by the growth of organisms (including zoogleal bacteria) in aeration tanks in the presence of dissolved oxygen. The term "activated" comes from the fact that the particles are teaming with bacteria, fungi, and protozoa.

Activated Sludge Process (ACK-ta-VATE-ed sluj): A biological wastewater treatment process in which a mixture of wastewater and activated sludge is aerated and agitated. The activated sludge is subsequently separated from the treated wastewater (mixed liquor) by sedimentation, and wasted or returned to the process as needed.

Adsorption (add-SORP-shun): To gather (a gas, liquid, or dissolved substance) on the surface or interface zone of another substance.

Advanced Waste Treatment: Any process of water renovation that upgrades water quality to meet specific reuse requirements. May include general cleanup of water or removal of specific parts of wastes insufficiently removed by conventional treatment processes.

Aeration Bay (air-A-shun): The same as aeration tank or aerator. The tank where raw or settled wastewater is mixed with return sludge and aerated.

Aeration Liquor: Mixed liquor. The contents of the aeration tank, which is composed of living organisms plus material carried into the tank by the untreated wastewater or primary effluent.

Aerobic (AIR-O-bick): A condition in which "free" or dissolved oxygen is present in the aquatic environment.

Aerobic Bacteria (AIR-O-bick back-TEAR-e-ah): Bacteria which live and reproduce only in an environment containing oxygen which is available for their respiration (breathing), such as atmospheric oxygen or oxygen dissolved in water. Oxygen combined chemically, such as in water molecules, H_2O, cannot be used for respiration by aerobic bacteria.

Aerobic Decomposition (AIR-O-bick): Decomposition and decay of organic material in the presence of "free" or dissolved oxygen.

---

*EPA Manual

Aerobic Process (AIR-O-bick): A waste treatment process conducted under aerobic (in the presence of "free" or dissolved oxygen) conditions.

Agglomeration (a-GLOM-er-A-shun): The growing or coming together of dispersed suspended matter into larger flocs or particles which settle rapidly.

Aliquot (AL-li-kwot): Portion of a sample.

Ambient Temperature (AM-bee-ent): Temperature of the surroundings.

Amperometric (am-PURR-o-MET-rick): A method of measurement that records electric current flowing or generated, rather than recording voltage. Amperometric titration is an electrometric means of measuring concentrations of substances in water.

Anaerobic (AN-air-O-bick): A condition in which "free" or dissolved oxygen is *not* present in the aquatic environment.

Anaerobic Bacteria (An-air-O-bick back-TEAR-e-ah): Bacteria that live and reproduce in an environment containing no "free" or dissolved oxygen. Anaerobic bacteria obtain their oxygen supply by breaking down chemical compounds which contain oxygen, such as sulfates (SO_4).

Anaerobic Decomposition (An-air-O-bick): Decomposition and decay of organic material in an environment containing no "free" or dissolved oxygen.

Anaerobic Digestion (An-air-O-bick): Wastewater solids and water (about 5% solids, 95% water) are placed in a large tank where bacteria decompose the solids in the absence of dissolved oxygen. At least two general groups of bacteria act in balance: (1) *Saprophytic* bacteria break down complex solids to volatile acids, and (2) *Methane Fermenters* break down the acids to methane, carbon dioxide, and water.

BOD (BEE-OH-DEE): See Biochemical Oxygen Demand.

BTU (BEE-TEA-YOU): British Thermal Unit. The amount of heat required to raise the temperature of one pound of water one degree Fahrenheit.

Bacteria (back-TEAR-e-ah): Bacteria are living organisms, microscopic in size, which consist of a single cell. Most bacteria utilize organic matter for their food and produce waste products as the result of their life processes.

Bacterial Culture (back-TEAR-e-al): In the case of activated sludge, the bacterial culture refers to the group of bacteria classed as *Aerobes*, and facultative organisms, which covers a wide range of organisms. Most treatment processes in the United States grow facultative organisms which utilize the carbonaceous (carbon compounds) BOD. Facultative organisms can live when oxygen resources are low. When "nitrification" is required, the nitrifying organisms are *Obligate Aerobes* (require oxygen) and must have at least 0.8 mg/1 of dissolved oxygen throughout the whole system to function properly.

Batch Process: A batch process is a treatment process in which a tank or reactor is filled, the water is treated, and the tank contents are released. The tank may then be filled and the process repeated.

Biochemical Oxygen Demand or BOD: The BOD indicates the rate of oxygen utilized by wastewater under controlled conditions of temperature and time.

Bioassay (BUY-o-ass-SAY): (1) an assay method using a change in biological activity as a qualitative or quantitative means of analyzing a material's response to biological treatment, or (2) A method of determining toxic effects of industrial wastes or other wastes by using live organisms such as fish for test organisms.

Biodegradation (BUY-o-de-grah-DAY-shun): The breakdown of organic matter by bacteria to more stable forms which will not create a nuisance or give off foul odors.

Bioflocculation (BUY-o-flock-u-LAY-shun): A condition whereby organic materials tend to be transferred from the dispersed form in wastewater to settleable material by mechanical entrapment and assimilation.

Blank: A bottle containing dilution water or distilled water, but the sample being tested is not added. Tests are frequently run on a *sample* and a *blank* and the differences compared.

Buffer: A measure of the ability or capacity of a solution or liquid to neutralize acids or bases. This is a measure of the capacity of water or wastewater for offering a resistance to changes in the pH.

Bulking (BULK-ing): Bulking occurs in activated sludge plants when the sludge becomes too light and will not settle properly.

Cathodic Protection (ca-THOD-ick): An electrical system for prevention of rust, corrosion, and pitting of steel and iron surfaces in contact with water and wastewater.

Chloramines (KLOR-a-means): Chloramines are compounds formed by the reaction of chlorine with ammonia.

Chlorine Demand: Chlorine demand is the difference between the amount of chlorine added to wastewater and the amount of residual chlorine remaining after a given contact time. Chlorine demand may change with dosage, time, temperature, pH, nature, and amount of the impurities in the water.

Chlorine Requirement: The amount of chlorine which must be added to produce the desired result under stated conditions. The result (the purpose of chlorination) may be based on any number of criteria, such as a stipulated coliform density, a specified residual chlorine concentration, the destruction of a chemical constituent, or others. In each case a definite chlorine dosage will be necessary. This dosage is the chlorine requirement.

Chlororganic (chlor-or-GAN-nick): Chlororganic compounds are organic compounds combined with chlorine. These compounds generally originate from or are associated with living or dead organic materials.

Clarifier (KLAIR-i-fire): Settling Tank, Sedimentation Basin. A tank or basin in which wastewater is held for a period of time, during which the heavier solids settle to the bottom and the lighter material will float to the water surface.

Coagulants (co-AGG-you--lents): Chemicals added to destabilize, aggregate, and bind together colloids and emulsions to improve settleability, filterability, or drainability.

Coliform (COAL-i-form): The coliform group of organisms is a bacterial indicator of contamination. This group has as one of its primary habitats the intestinal tract of human beings. Coliforms also may be found in the intestinal tract of warm-blooded animals, and in plants, soil, air, and the aquatic environment.

Colloids (KOL-loids): Very small solids (particulate or insoluble material) in finely divided form that remain dispersed in a liquid for a long time due to their small size and electrical charge.

Colorimetric: A means of measuring unknown concentrations of water quality indicators in a sample by comparing the sample's color, after the addition of specific reagents, with the color of known concentrations.

Combined Sewer: A sewer designed to carry both sanitary wastewaters and storm or surface water runoff.

Communution (com-min-OO-shun): A mechanical treatment process which cuts large pieces of wastes into smaller pieces so they won't plug pipes or damage equipment (shredding).

Comminutor (com-min-OO-ter): A device used to reduce the size of the solid chunks in wastewater by shredding (comminuting). The shredding action can be visualized if you imagine many scissors cutting or hammering to shreds all the large influent solids material.

Composite (proportional) Samples (com-POZ-it): Samples collected at regular intervals in proportion to the existing flow and then combined to form a sample respresentative of the entire period of flow over a given period of time.

Coning (CONE-ing): A condition that may be established in a sludge hopper during sludge withdrawal when part of the sludge moves toward the outlet while the remainder tends to stay in place. Development of a cone or channel of moving liquid surrounded by relatively stationary sludge.

Conventional Treatment: The pretreatment, sedimentation, flotation, trickling filter and activated sludge wastewater treatment processes.

Cross-Connection: A connection where wastewater or water from a pump seal could enter a drinking water supply.

DO (DEE-OH): Abbreviation of Dissolved Oxygen. DO is the atmospheric oxygen dissolved in water or wastewater.

Dateometer (date-O-meter): A small calendar disc attached to motors and equipment to indicate the year in which the last maintenance service was performed.

Decomposition, Decay: Generally aerobic processes that convert unstable materials into more stable forms by chemical or biological action. Waste treatment encourages decay in a controlled situation in order that the material may be disposed of in a stable form. When organic matter decays under anaerobic conditions (putrefaction), undesirable odors are produced. In aerobic processes, the odors are much less objectionable than those produced by anaerobic decomposition.

Degradation (de-grah-DAY-shun): The conversion of a substance to simpler compounds.

Density (DEN-sit-tee): The weight per unit volume of any substance. The density of water (at 4°C) is 1.0 gram per cubic centimeter (gms/cc) or about 62.4 lbs per cubic foot.

Detention Time: The time required to fill a tank at a given flow or the theoretical time required for a given flow of wastewater to pass through a tank.

Detritus (de-TRI-tus): The heavy, course material carried by wastewater.

Dewaterable: A material is considered dewaterable if water will readily drain from it. Generally raw sludge dewatering is more difficult than water removal from digested sludge.

Diffused Air Aeration: A diffused air activated sludge plant takes air, compresses it, and then discharges the air below the water surface of the aerator through some type of air diffusion device.

Diffuser: A diffuser is a device (porous plant, tube, bag) used to break the air stream from the blower system into fine bubbles in the mixed liquor.

Digester (die-JEST-er): A tank in which sludge is placed to allow sludge digestion to occur. Digestion may occur under anaerobic (more common) or aerobic conditions.

Disinfection (DIS-in-feck-shun): The process by which pathogenic (disease) organisms are killed. There are several ways to disinfect, but chlorination is the most frequently used method in water and wastewater treatment.

Dissolved Oxygen: Atmospheric oxygen dissolved in water or wastewater, usually abbreviated DO.

Distillate: In the distillation of a sample, a portion is evaporated; the part that is condensed afterwards is the distillate.

Distributor: The rotating mechanism that distributes the wastewater evenly over the surface of a trickling filter or other process unit. Also see Fixed Spray Nozzle.

Effluent (EF-lu-ent): Wastewater or other liquid — raw, partially or completely treated — flowing *from* a basin, treatment process, or treatment plant.

Elutriation (e-LOO-tree-a-shun): The washing of digested sludge in plant effluent with a suitable ratio of sludge to effluent. The objective is to remove (wash out) fine particulates or certain soluble components in sludge.

Emulsion (e-MULL-shun): A liquid mixture of two or more liquid substances not normally dissolved in one another, but one liquid held in suspension in the other.

End Point: Samples are titrated to the end point. This means that a chemical is added, drop by drop, to a sample until a certain color change (blue to clear, for example) occurs which is called the *end point* of the titration. In addition to a color change, an end point may be reached by the formation of a precipitate or the reaching of a specified pH. An end point may be detected by the use of an electronic device such as a pH meter.

Endogenous (en-DODGE-en-us): A diminished level of respiration in which materials previously stored by the cell are oxidized.

Enteric: Intestinal.

Enzymes (EN-zimes): Enzymes are substances produced by living organisms that speed up chemical changes.

Estuaries (ES-chew-wer-eez): Bodies of water at the lower end of a river that are subject to tidal fluctuations.

Facultative (FACK-ul-tay-tive): Facultative bacteria can use either molecular (dissolved) oxygen or oxygen obtained from food materials. In other words, facultative bacteria can live under aerobic or anaerobic conditions.

Facultative Pond (FACK-ul-tay-tive): The most common type of pond in current use. The upper portion (supernatant) is aerobic, while the bottom layer is anaerobic. Algae supply most of the oxygen to the supernatant.

Filamentous Bacteria (FILL-a-men-tuss): Organisms that grow in a thread of filamentous form.

Fixed: A sample is "fixed" in the field by adding chemicals that prevent the water quality of the sample from changing before final measurements are performed later in the lab.

Fixed Spray Nozzle: Cone-shaped spray nozzle used to distribute wastewater over the filter media, similar to a lawn sprinkling system. A deflector or steel ball is mounted within the cone to spread the flow of wastewater through the cone, causing a spraying action. Also see Distributor.

Flame Polished: Sharp or broken edges of glass (such as the end of a glass tube) are flame polished by placing the edge in a flame and rotating it. By allowing the edge to melt slightly, it will become smooth.

Flights: Scraper boards, made from redwood or other rot-resistant woods, used to collect and move settled sludge or floating scum.

Floc: Groups or "clumps" of bacteria that have come together and formed a cluster. Found in aeration tanks and secondary clarifiers.

Flocculated (FLOCK-you-lay-ted): An action resulting in the gathering of fine particles to form larger particles.

Freeboard: The vertical distance from the normal water surface to the top of the confining wall.

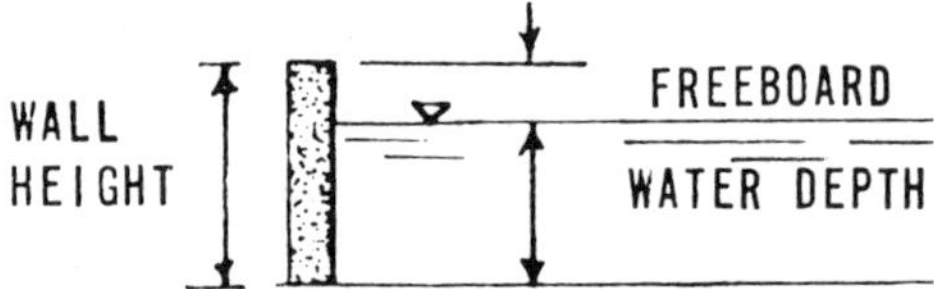

Grit: The heavy mineral material present in wastewater such as sand, eggshells, gravel, and cinders.

Grit Removal : Grit removal is accomplished by providing an enlarged channel which causes the flow velocity to be reduced and allows the heavier grit to settle to the bottom of the channel where it can be removed.

Head Loss: "Head" is a common term used in discussing pumps. It is a way of expressing pressure in terms of the height of a vertical column of water. In the sketch, the head loss is the height to which the water must build up until there is sufficient pressure to force that particular amount of water through the slots in the comminutor drum.

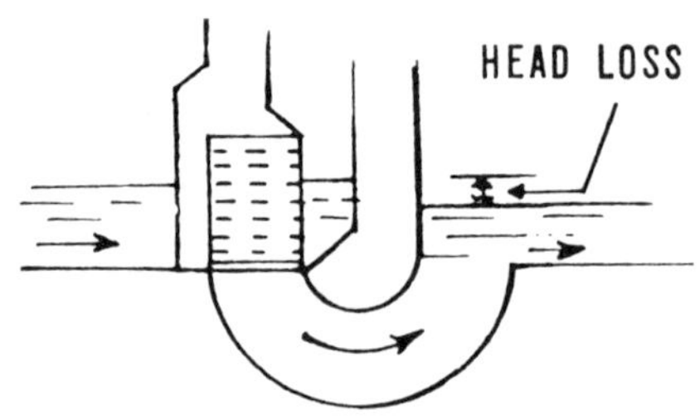

Hepatitis: Hepatitis is an acute viral infection of the liver (yellow jaundice).

Hydrolysis (hi-DROL-e-sis): The addition of water to the molecule to break down complex substances into simpler ones.

Hypochlorinators (hi-po-KLOR-i-NAY-tors): Hypochlorinators are devices that are used to feed calcium, sodium, or lithium hypochlorite as the disinfecting agent.

Hypochlorites (hi-po-KLOR-ites): Hypochlorites are compounds containing chlorine that are used for disinfection. They are available as liquids or solids (powder, granules, and pellets) in barrels, drums, and cans.

Imhoff Cone: A clear, cone-shaped container marked with graduations used to measure the volumetric concentration of settleable solids in wastewater.

Infiltration (In-fill-TRAY-shun): Groundwater that seeps into pipes through cracks, joints, or breaks.

Influent (IN-flu-ent): Wastewater or other liquid — raw or partially treated — flowing *into* a reservoir, basin, treatment process, or treatment plant.

Inoculate (in-NOCK-you-LATE): To introduce a seed culture into a system.

Inorganic Waste: Waste material such as sand, salt, iron, calcium, and other mineral materials which are not converted in large quantities by organism action. Inorganic wastes are chemical substances of mineral origin and may contain carbon and oxygen, whereas organic wastes are chemical substances of animal or vegetable origin and contain mainly carbon and hydrogen along with other elements.

Launders (LAWN-ders): Sedimentation tank effluent troughs.

Lineal (LIN-e-al): The length in one direction of a line. For example, a board 12 feet long has 12 lineal feet in its length.

Liquefaction (LICK-we-FACK-shun): Liquefaction as applied to sludge digestion means the transformation of large solid particles of sludge into either a soluble or a finely dispersed state.

Loading : Quantity of material applied to a device at one time.

M or Molar: A molar solution consists of one gram molecular weight of a compound dissolved in enough water to make one liter of solution. A gram molecular weight is the molecular weight of a compound in grams. For example, the molecular weight of sulfuric acid (H_2SO_4) is 98. A 1 M solution of sulfuric acid would consist of 98 grams of H_2SO_4 dissolved in enough distilled water to make one liter of solution.

MPN (EM-PEA-EN): MPN is the Most Probable Number of coliform group organisms per unit volume expressed as a density of organisms per 100 ml.

Manometer (man-NOM-meet-her): Usually a glass tube filled with a liquid and used to measure the difference in pressure across a flow measuring device such as an orifice or venturi meter.

Masking Agents: Liquids which are dripped into the wastewater, sprayed into the air, or evaporated (using heat) with the "fumes" or odors discharged into the air by blowers to make an undesirable odor less noticeable.

Mechanical Aeration: The surface of the aeration tank is agitated to cause spray and waves by a paddle wheel, mixers, rotating brushes, pumps discharging water into the air like a fountain discharging the water down a series of steps creating falls or some other method of splashing water into the air or air into the water where the oxygen can be absorbed.

Media: The material in a trickling filter over which settled wastewater is sprinkled and then flows over and around during treatment. Slime organisms grow on the surface of the media and treat the wastewater.

Meniscus: The curved top of a column of liquid (water, oil, mercury) in a small tube. Water will form a valley when the liquid wets the walls of the tube, while mercury will form a hill and the walls of the tube are not wetted.

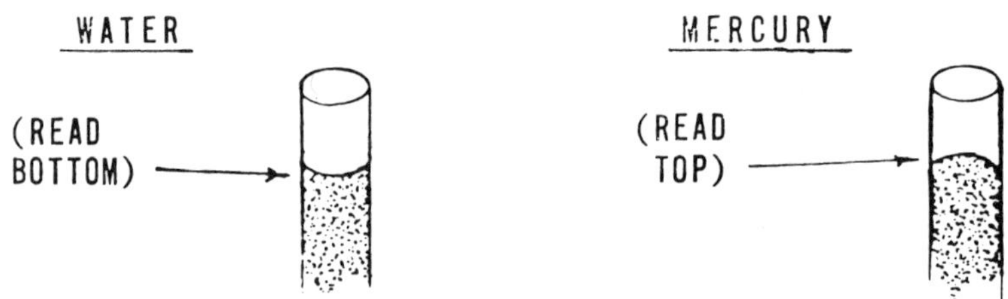

Mesophilic Bacteria (mess-O-FILL-lick): Medium temperature: A group of bacteria that thrive in a temperature range between 68°F and 113°F.

Microorganisms (micro-ORGAN-is-zums): Very small organisms that can be seen only through a microscope. Some microorganisms use the wastes in wastewater for food and thus remove or alter much of the undesirable matter.

Milligrams Per Liter, mg/1 (MILL-i-GRAMS per LEET-er): A measure of the concentration by weight of a substance per unit volume. For practical purposes, one mg/1 is equal to one part per million parts (ppm). Thus a liter of water with a specific gravity of 1.0 weighs one million milligrams and if it contains 10 milligrams of dissolved oxygen, the concentration is 10 milligrams per million milligrams, or 10 milligrams per liter (10 mg/1), or 10 parts of oxygen per million parts of water, or 10 parts per million (10 ppm).

Millimicron (MILL-e-MY-cron): One thousandth of a micron or a millionth of a millimeter.

Mixed Liquor: When the activated sludge in an aeration tank is mixed with primary effluent or the raw wastewater and return sludge, this mixture is then referred to as mixed liquor as long as it is in the aeration tank. When the mixed liquor flows from the aeration tank it goes into the secondary clarifiers or final sedimentation tank. Mixed liquor also may refer to the contents of mixed aerobic or anaerobic digesters.

Molecular Weight: The molecular weight of a compound in grams is the sum of the atomic weights of the elements in the compound. The molecular weight of sulfuric acid (H_2SO_4) in grams is 98.

Element	Atomic Weight	Number of Atoms	Molecular Weight
H	1	2	2
S	32	1	32
O	16	4	64
			98

Molecule (MOLL-ee-kule): The smallest portion of an element or compound retaining or exhibiting all the properties of the substance.

Motile (MO-till): Motile organisms exhibit or are capable of movement.

196

Muffle Furnace: A small oven capable of temperatures up to 600°C and used in laboratories for burning or incinerating samples to determine their loss on ignition (volatile) or fixed solids (ash) content.

Multi-Stage Pump: A pump that has more than one impeller. A single-stage pump has one impeller.

N or Normal: A normal solution contains one gram equivalent weight or a reactant (compound) per liter of solution. The equivalent weight of an acid is that weight of which contains one gram atom of ionizable hydrogen or its chemical equivalent. For example, the equivalent weight of sulfuric acid (H_2SO_4) is 49 (98 divided by 2 because there are two replaceable hydrogen ions). A 1 N solution of sulfuric acid would consist of 49 grams of H_2SO_4 dissolved in enough water to make one liter of solution.

Nitrification: The biochemical conversion of unoxidized nitrogenous matter (ammonia and organic nitrogen) to oxidized nitrogen (usually nitrate). The second-stage BOD is sometimes referred to as the nitrification stage (first-stage BOD is called the carbonaceous stage — carbon compounds oxidized to CO_2).

Nomogram: A chart or diagram containing three or more scales used to solve problems with three or more variables instead of using mathematical formulas.

Nonsparking Tools: These tools will not produce a spark during use.

Nutrients: Substances which are required to support living plants and organisms. Major nutrients are carbon, hydrogen, oxygen, sulfur, nitrogen and phosphorus. Nitrogen and phosphorus are difficult to remove from wastewater by conventional treatment processes because they are water soluble and tend to recycle.

Obligate Aerobes: Bacteria that must have molecular (dissolved) oxygen (DO) to survive.

Organic Waste: Waste material which comes from animal or vegetable sources. Organic waste generally can be consumed by bacteria and other small organisms. Inorganic wastes are chemical substances of mineral origin and may contain carbon and oxygen, whereas organic wastes contain mainly carbon and hydrogen along with other elements.

Orifice (OR-i-fiss): An opening in a plate, wall, or partition. In a trickling filter distributor the wastewater passes through an orifice to the surface of the filter media. An orifice flange set in a pipe consists of a slot or hole smaller than the pipe diameter. The difference in pressure in the pipe above and below the orifice may be related to flow in the pipe.

Orthotolidine (or-tho-TOL-i-dine): Orthotolidine is a colorimetric indicator of chlorine residual in which a yellow-colored compound is produced.

Oxidation (ox-i-DAY-shun): Oxidation is the addition of oxygen, removal of hydrogen, or the removal of electrons from an element or compound. In wastewater treatment organic matter is oxidized to more stable substances.

Parasitic Bacteria (PARA-SIT-tick): Parasitic bacteria are those bacteria which normally live off another living organism, known as the host.

Pathogenic Organsism (path-o-JEN-nick OR-gan-iz-ums): Bacteria or viruses which can cause disease (typhoid, cholera, dysentery). There are many types of bacteria which do not cause disease and which are *not* called pathogenic. Many beneficial bacteria are found in wastewater treatment processes actively cleaning up organic wastes.

Percent Saturation: Liquids can contain in solution limited amounts of compounds and elements. 100% saturation is the maximum theoretical amount that can be dissolved in the solution. If more than the maximum theoretical amount is present, the solution is supersaturated.

$$\text{Percent Saturation} = \frac{\text{Amount in Solution}}{\substack{\text{Maximum Theoretical} \\ \text{Amount in Solution}}} \times 100\%$$

pH (PEA-A-ch): pH is an expression of the intensity of the alkaline or acidic strength of a water. Mathematically, pH is the logarithm (base 10) of the reciprocal of the hydrogen ion concentration.

$$pH = \text{Log} \ \frac{1}{(H^+)}$$

The pH may range from 0 to 14, where 0 is most acid, 14 most alkaline, and 7 neutral. Natural waters usually have a pH between 6.5 and 8.5

Photosynthesis (foto-SIN-the-sis): A process in which organisms with the aid of chlorophyll (green plant enzyme) convert carbon dioxide and inorganic substances to oxygen and additional plant material, utilizing sunlight for energy. Land plants grow by the same process.

Physical Waste Treatment Processes: Racks, screens, comminutors, sedimentation, and flotation. Chemical or biological reactions are not an important part of the process.

Pollution: Any interference with beneficial reuse of water or failure to meet quality requirements.

Ponding: A condition occurring on trickling filters when the voids become plugged to the extent that water passage through the filter is inadequate. Ponding may be the result of excessive slime growths, trash, or media breakdown.

Population Equivalent: A means of expressing the strength of organic material in wastewater. Domestic wastewater consumes, on an average, approximately 0.2 lb of oxygen per person per day, as measured by the standard BOD test.

Postchlorination: Chlorination of the plant discharge or effluent *following* plant treatment.

Preaeration: A preparatory treatment of wastewater consisting of aeration to freshen the wastewater, remove gases, add oxygen, promote flotation of grease, and aid coagulation.

Prechlorination: Chlorination at the headworks of the plant; influent chlorination *prior* to plant treatment.

Pretreatment: Use of racks, screens, comminutors, and grit removal devices to remove metal, rocks, sand, eggshells, and similar materials which may hinder operation of a treatment plant.

Primary Treatment: A wastewater treatment process consisting of a rectangular or circular tank which allows those substances in wastewater that readily settle or float to be separated from the water being treated.

Protozoa (pro-toe-ZOE-ah): A group of microscopic animals, principally of one cell, that sometimes cluster into colonies.

Prussian Blue: A paste or liquid used to show a contact area.

Psychrophilic Bacteria (sy-kro-FILL-ik): Cold Temperature: A group of bacteria that thrive in temperatures below $68°F$.

Putrefaction (PU-tree-FACK-shun): Biological decomposition of organic matter with the production of ill-smelling products associated with anaerobic conditions.

Putrescible (pu-TRES-sib-bull): Putrescible material will decompose under anaerobic conditions and produce nuisance odors.

Rack: Parallel metal bars or rods evenly spaced and placed at an angle in the influent channel that remove rags, rocks, and cans from wastewater.

Raw Wastewater: Plant influent or wastewater before any treatment.

Reagent (re-A-gent): A substance which takes part in a chemical reaction that is used to measure, detect, or examine other substances.

Receiving Water: A stream, river, lake, or ocean into which treated or untreated wastewater is discharged.

Recirculation: The return of part of the effluent from a treatment process to the incoming flow.

Reliquifaction (re-LICK-we-FACK-shun): The return of gas to a liquid. For example, a condensation of chlorine gas returning to the liquid form.

Representative Sample : A portion of material or water identical in content to that in the larger body of material or water being sampled.

Residual Chlorine: Residual chlorine is the amount of chlorine remaining after a given contact time and under specified conditions.

Respiration: The physical and chemical processes by which an organism supplies its cells and tissues with oxygen needed for metabolism and relieves them of carbon dioxide formed in energy-producing reactions.

Rising Sludge: Rising sludge occurs in the secondary clarifiers of activated sludge plants when the sludge settles to the bottom of the clarifier, is compacted, and then starts to rise to the surface.

Sanitary Sewer (SAN-eh-tare-ee SUE-er): A sewer intended to carry wastewater from homes, business, and industries. Storm water runoff sometimes is collected and transported in a separate system of pipes.

Saprophytic Organisms (SAP-pro-FIT-tik): Organisms living on dead or decaying organic matter. They help natural decomposition of the organic solids in wastewater.

Screen: A device with openings generally uniformly sized to retain or remove suspended or floating objects in wastewater larger than the openings. A screen may consist of bars, rods, wires, gratings, wire mesh, or perforated plates.

Secondary Treatment: A wastewater treatment process used to convert dissolved or suspended materials into a form more readily separated from the water being treated.

Septic (SEP-tick): A condition produced by the growth of anaerobic organisms. If severe, the wastewater turns black, giving off foul odors and creating a heavy oxygen demand.

Septicity (sep-TIS-it-tee): Septicity is the condition in which organic matter decomposes to form foul-smelling products associated with the absence of free oxygen.

Sewage: The used water and solids from homes that flow to a treatment plant. The preferred term is wastewater.

Shock Load: The arrival at a plant of a waste which is toxic to organisms in sufficient quantity or strength to cause operating problems, such as odors or sloughing off of the growth or slime on the trickling filter media. Organic or hydraulic overloads also can cause a shock load.

Shredding: A mechanical treatment process which cuts large pieces of wastes into smaller pieces so they won't plug pipes or damage equipment (comminution).

Sloughings (SLUFF-ings): Trickling filter slimes that have been washed off the filter media. They are generally quite high in BOD and will degrade effluent quality unless removed.

Sludge (sluj): The settleable solids separated from liquids during processing or deposits on bottoms of streams or other bodies of water.

Sludge Digestion: A process by which organic matter in sludge is gasified, liquefied, mineralized, or converted to a more stable form by anaerobic (more common) or aerobic organisms.

Sludge Gasification: A process in which soluble and suspended organic matter are converted into gas. Sludge gasification will form bubbles of gas in the sludge and cause large clumps of sludge to rise and float on the water surface.

Specific Gravity: Weight of a particle or substance in relation to the weight of water. Water has a specific gravity of 1.000 at 4°C (or 39°F). Wastewater particles usually have a specific gravity of 0.8 to 2.6.

Stabilize: To convert to a form that resists change. Organic material is stabilized by bacteria which convert the material to gases and other relatively inert substances. Stabilized organic material generally will not give off obnoxious odors.

Stabilized Waste: A waste that has been treated or decomposed to the extent that, if discharged or released, its rate and state of decomposition would be such that the waste would not cause a nuisance or odors.

Stasis (STAY-sis): Stagnation or inactivity of the life processes within organisms.

Stethoscope: An instrument used to magnify sounds and convey them to the ear.

Storm Sewer: A separate sewer that carries runoff from storms, surface drainage, and street wash, but excludes domestic and industrial wastes.

Stuck: A stuck digester does not decompose organic matter properly. It is characterized by low gas production, high volatile acid to alkalinity relationship, and poor liquid-solids separation. A digester in a stuck condition is sometimes called a "sour" digester.

Supernatant (sue-per-NAY-tent): Liquid removed from settled sludge. Supernatant commonly refers to the liquid between the sludge on the bottom and the scum on the surface of an anaerobic digester. This liquid is usually returned to the influent wet well or the primary clarifier.

Tertiary Treatment (TER-she-AIR-ee): See Advanced Waste Treatment.

Thermophilic Bacteria (thermo-FILL-lik): Hot temperature: A group of bacteria that thrive in temperatures above 113°F.

Thief Hole: A digester sampling well.

Titrate: To *titrate* a sample, a chemical solution of known strength is added on a drop-by-drop basis until a color change, precipitate, or pH change in the sample is observed (end point). Titration is the process of adding the chemical solution to completion of the reaction as signaled by the end point.

Totalizer: A totalizer continuously sums or adds up the flow into a plant in gallons or million gallons or some other unit of measurement.

Toxicity (tocks-IS-it-tee): A condition that may exist in wastes that will inhibit or destroy the growth or function of any organism.

Trickling Filter: A treatment process in which the wastewater trickles over media that provide the opportunity for the formation of slimes which clarify and oxidize the wastewater.

Trickling Filter Media: Rocks or other durable materials that make up the body of the filter. Synthetic (manufactured) media have been used successfully.

Two-State Filters: Two filters are used. Effluent from the first filter goes to the second filter, either directly or with a clarifier between the two filters.

Volute (vol-LOOT): The spiral-shaped casing surrounding a pump impeller that collects the liquid discharged by the impeller.

Wastewater: The used water and solids from a community that flow to a treatment plant. Storm water, surface water, and groundwater infiltration also may be included in the wastewater that enters a plant. The term sewage usually refers to household wastes, but this word is being replaced by the term wastewater.

Weir (weer): A vertical obstruction, such as a wall, or plant, placed in an open channel and calibrated in order that a depth of flow over the weir can easily be converted to a flow rate in MGD (million gallons per day).

Weir Diameter (weer): Circular clarifiers have a circular weir within the outside edge of the clarifier, and all of the water leaving the clarifier flows over this weir to the opposite edge and passing through the center of the circle formed by the weir.

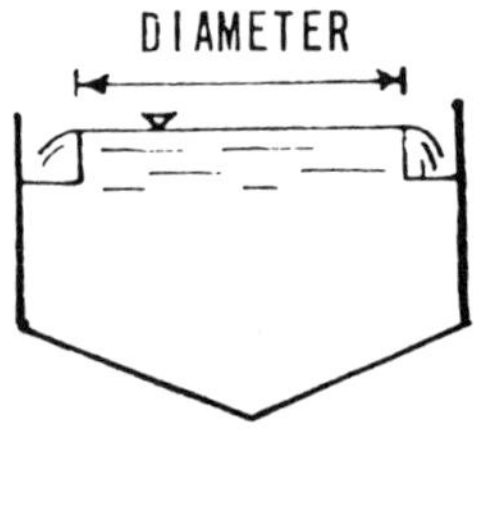

Weir, Proportional (weer): A specially shaped weir in which the flow through the weir is directly proportional to the head.

Wet Oxidation: Any process in which substances are converted to a higher oxidation state in a water media, such as activated sludge, trickling filters, ponds, or digesters.

202

Zoogleal Film (ZOE-glee-al): A complex population of organisms that form a slime growth on the trickling filter media and break down the organic matter in wastewater. These slimes consist of living organisms feeding on the wastes in wastewater, dead organisms, silt, and other debris. Slime growth is a more common word.

Zoogleal Mass (ZOE-glee-al): Jelly-like masses of bacteria found in both the trickling filter and activated sludge processes. These masses may be formed for or function as the protection against predators and for storage of food supplies.

APPENDIX B

Suggested Lubrication Materials

Worm Gear Reducers

Ambient Temp	Max. Operating Temp	Type
$-30°$ to $15°$F	$150°$F	Vactra oil #2
$16°$ to $50°$F	$185°$F	Mobil Compound #DD
$51°$ to $110°$F	$225°$F	Mobil Cyl. Oil #600W
$111°$ to $165°$F	$225°$F	Mobil Cyl. Oil #600W

Helical Gear Reducers

Ambient Temp	Max. Operating Temp	Type
$-30°$ to $15°$F	$150°$F	Vactra oil #2
$16°$ to $50°$F	$185°$F	Mobil Compound #AA
$51°$ to $110°$F	$225°$F	Mobil Compound #BB
$111°$ to $165°$F	$225°$F	Mobil Compound #600W

Note #1 — The operating temperature is approximately $120° - 130°$F above ambient temperature.

Note #2 — Mobil Oil Co. was specified above only because of its wide distribution. Other companies provide equivalent products. All major oil companies offer the needed lubricants under different names. It is recommended that the operator contact a major supplier whose "on the spot" expertise is available.

APPENDIX C

PUMPS FOR POLLUTION CONTROL

by James A. Edwards, Field Editor, Pollution Engineering

Pumps are the heart of a waste treatment system. They serve a multitude of purposes, and are available in a wide range of designs that can be adapted to any need. Designs include the familiar centrifugal so frequently used for transferring solutions between reaction tanks, proportioning pumps for accurate control of chemical addition, diaphragm pumps for slurries, screw pumps for high viscosity solutions and gear pumps for hydraulic systems.

A typical system design is shown in Figure 1, incorporating centrifugal, proportioning and slurry pumps. Pump Number 1 is a centrifugal, which could have a wide range of characteristics. Pump Number 2 is a proportioning pump and can be either a diaphragm or piston type. Pump Number 3 is a slurry pump and is used for sludge transfer.

Fig. 1 Schematic of pumping system.

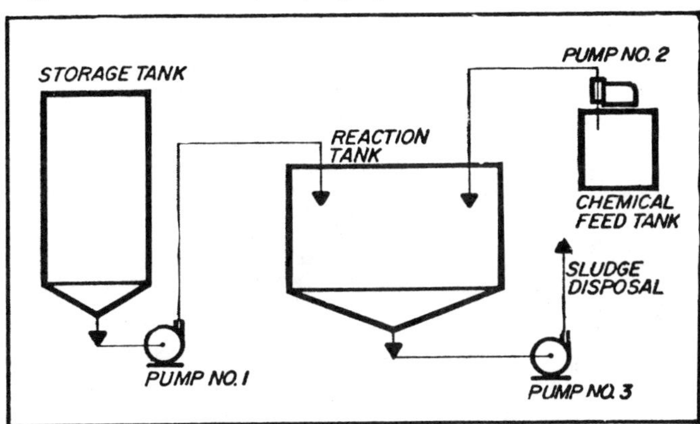

Let us consider the centrifugal pumps first. These are most commonly used in the waste treatment industry for liquid movement. Centrifugal pumps are available in a wide range of designs ranging from low discharge head, high volume pump to a high discharge head, high volume pump. Materials of construction are available to resist any chemical, and either non-priming or self-priming models can be purchased. Considering all of these variables, several factors must be considered in selecting a centrifugal pump.

1. Total discharge head (TDH),
2. Net positive suction head (NPHS),
3. Motor RPM,
4. Seal design,
5. Equipment location,
6. Materials of construction,
7. Solution temperature, and
8. Equipment application.

Table 1 lists materials of construction that are commercially available. Once a manufacturer has been selected who can meet the corrosion resistance requirements, the technical aspects of pump specification come into play. Referring again to Figure 1, consider Pump Number 1. This is a high volume, low head pump, transferring 300 gpm of solution from a reaction tank to a neutralization tank. The solution contains chromic acid, sulfuric acid and a low percentage of abrasive solids. The temperature is 100F and the solution pH is 1.5 From tables of corrosion data we find that available materials of construction are Hastelloy A, Duriron and 304 SS. Others such as tantalum are available but are not considered practical because of cost. Polyethylene, PVC, hypalon-lined and rubber-lined pumps are also applicable. However, if the temperatures were higher these would also be eliminated.

Table I Standard Materials of Construction for Pumps

Centrifugal	Stainless Steel (All Grades) Hastelloy A Hastelloy B Duriron Steel Brass Polypropylene Teflon lined Rubber lined Hypalon lined
Proportioning	Stainless Steel (All Grades) Steel Polypropylene
Diaphragm	Steel (Housings) Synthetic Rubbers Hypalon
Progressive Cavity	Synthetic Rubber (Stator) Hardened Steel (Rotor)

Stainless steel is not always a good selection. When stainless steel is used, definite grades must be specified, taking into consideration the temperature of the solution and the chemical concentration. An example of the problems that can be encountered in the use of 316 stainless steel with solutions containing hydrocholoric acid. This solution rapidly attacks the metal causing severe pitting, quickly destroying the piece of equipment. Cast iron is traditionally used for handling 98 percent sulfuric acid, yet this material will quickly dissolve at lower concentrations. Fiber-reinforced plastics (FRP) are not a panacea either. They have definite good points, but the properties of the plastic base compound must be considered when writing an equipment specification.

If both user and equipment suppliers are unable to decide on a usable material for pumps, a laboratory should be hired to evaluate the materials under consideration for corrosion resistance to the solutions to be handled. Ideally, these corrosion tests would be performed using solution samples provided by the user. However, if this is not possible, the following information should be specified for the tests:

1. Composition of the liquid.
2. Temperature of the liquid, and
3. Process steps just prior to this one (was the liquid aerated?).

The report from the laboratory will allow a final decision on materials of construction. In our case, with Pump Number 1, we have selected 304 stainless steel. Before we can call the pump manufacturer and request his best pump at the lowest price, some information on the operating characteristics to be met by the pump must be developed. The pumping rate is dependent on a multitude of factors, primarily pressure, motor RPM, impeller size and line size.

At this point, we will consider the three factors which the buyer must specify to be certain he is obtaining a pump that will meet his needs. These are the total discharge head (TDH), the net positive suction head (NPSH), and the total head. The total head is the head which the pump must move the liquid. It includes the net lift accomplished by the pump and the friction losses generated by the fluid moving through the piping. It is important to note emphasis on net lift by the pump. If the pump barely has a flooded suction, the static pressure on the pump suction is negligible. However, if the pump is installed on a tank which has a static head of 20 ft, the liquid can be lifted to a height of 20 ft without a pump. In calculating the total head, this 20 ft is subtracted from the total head. Finally, in calculating the total head, we must consider the line losses; that is, the resistance to flow generated by the friction of the fluid against the pipe wall. The size of the pump piping can severely restrict pump performance. Properly sized piping will enhance pump performance and reduce operating difficulties.

The TDH can be summarized as the following equation:

$$\textbf{TDH} = \textbf{total lift} + \textbf{line loss}$$

Another design term that is useful is the total head.

$$\textbf{Total head} = \textbf{discharge head} \pm \textbf{suction head}$$

In this equation, the two terms are added when the suction lift is negative.

Determination of the NPSH becomes more complex where the units are feet of liquid being pumped. This is the net head on the pump suction. It must be specified to make certain that cavitation will not take place, and to ensure that the designated pump can move the desired volume of liquid under the system operating conditions. NPSH is defined as follows:

$$\text{NPSH} = \text{atmospheric pressure} + \text{static head}$$
$$-\text{liquid vapor pressure} - \text{suction line loss}$$

As can be seen, when pumping a cold solution with almost no static head and properly sized piping, the NPSH is equal to the atmospheric pressure. Few manufacturers provide information on the NPSH range of their equipment in their literature, and this is a point that must be developed in conversations with them. If, for example, a manufacturer's pump cannot meet specified NPSH and this is the pump the buyer wants to use, what changes can be made in the system design to permit its use? Referring back to the equation, the following options are available:

1. **Change system design to increase static head on the pump** — This does not mean adding tankage. Quite often the system layout is such that the suction pipe can be a long vertical drop which will provide all the head needed when kept full.

2. **Change the piping size** — Increasing the suction line from 1-1/2 inches to 3 inches with a flow of 90 gpm will decrease the line loss by 10.6 psi for an equivalent length of 50 ft. This change increases system cost, but will probably be cheaper than purchasing a pump capable of operating under the low head conditions.

3. **Cool the solution** — A low NPSH is usually encountered when trying to transfer a hot liquid. Cooling the solution will lower the vapor pressure, increasing the NPSH. Using water as an example, lowering the temperature from 180F to 120F will drop the vapor pressure 5.8 psi and change the NPSH in our previous example by 13.4 ft.

4. **Pressurize the tank** — This has the same effect as raising the atmospheric pressure and will directly increase the NPSH.

   Now that the operating conditions have been specified, other options must be determined including motor RPM, seal design and impeller design.

   Motor RPM is important. Standard speeds are 1750 and 3500 rpm. The higher rating provides a greater discharge head and a larger volume handling capability in a smaller pump, this reducing the initial investment, especially where expensive materials of construction are involved.

Seals

Seal design plays an important role in pump maintenance. Two types of seals are used: the packed stuffing box and a mechanical seal. The packed gland has several layers of packing placed inside a compression ring, which is gradually tightened as the packing wears away. This type of seal is fine for use on clear, non-hazardous solutions. It should not be used for slurries or solutions that are corrosive. Properly used, a packed stuffing box provides an inexpensive low-maintenance pump seal. When improperly applied, it is a constant headache. This seal depends on the solution being pumped for lubrication; the compression ring should never be tightened to the point where all flow is stopped. Rather, a slow drip must be maintained out of the stuffing box. This prevents damage to the packing. However, if the solution contains suspended solids or is a slurry, the packing is rapidly eroded and a severe leak develops.

A mechanical seal is completely different in its approach. It consists of one stationary face and one rotating member held together by springs, a backup plate and other mechanisms. These designs provide a true mechanical seal, which prevents the solution being pumped from leaking out of the housing. An option to be considered when handling slurries or solutions that might crystallize is a mechanical seal with a flushed face. In this design, tap water is fed between the two seal faces to keep them clean, preventing them from scoring and increasing the life of the seal. This last item, seal life, is very important. Mechanical seals are expensive and constant replacement will quickly run up maintenance cost. Spare parts for one mechanical seal can easily cost in excess of $130 depending on the pump, shaft size, etc.

Another pump design is one that uses no seals. Typical of these are the magnetic drive pumps and those which use expeller blades to develop a pressure greater than the intake head. This latter design utilizes centrifugal force, while the pump is in motion, to remove liquid which slips behind the impeller, by adding it to the liquid leg in the expeller. In effect the centrifugal force created by the expeller is a seal, which can replace the other designs discussed.

Impellers

There are several types of impeller design, Figure 2. The open and semi-enclosed designs are normally used for small pumps, or pumps handling scaling liquids or liquids containing coarse solids. In other cases, the shrouded impeller is used, since the design provides higher efficiencies and performance is less affected by wear or erosion.

It is not wise to buy a pump with a large housing and trim the impeller to a smaller size. For maximum efficiency, always try to utilize the largest impeller possible for a given pump. The closer the tip of the impeller is to the edge of the housing, the more efficiently the energy put into the pump will be used.

In addition to the standard horizontal centrifugal pumps which have been discussed, a wide choice of vertical centrifugal pumps is available. These have the same design characteristics discussed above, but are applied differently. The familiar sump pump is the most obvious example of the effective application of these pumps. If the solution to be transferred is collected in a sump or pit, a vertical

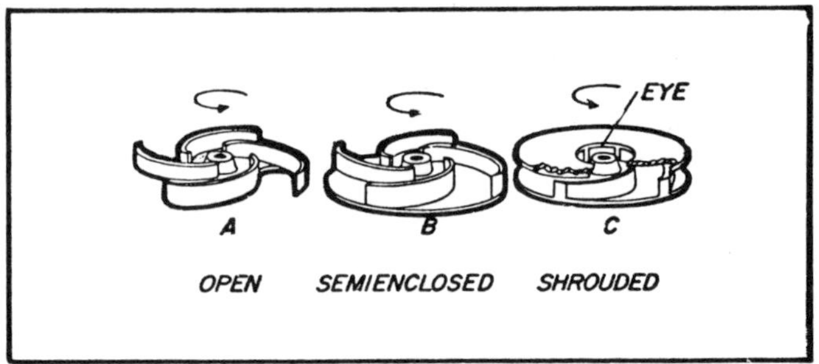

Fig. 2 Standard impeller designs.

pump can be used for the transfer step. Depending upon the head requirements and the solution to be transferred, single or multistage pumps are available in a wide choice of materials, including rubber-lined which can be used for transfering light sludges or slurries.

A variation of the vertical pump is the single or multistage vertical turbine pump developed from the deepwell pump. They are designed to operate at a low NPSH, moving large volumes of water against relatively high heads. These pumps have been designed for both submersible and non-submersible applications, permitting their use in a variety of applications.

Chemical Pumps

The pump most commonly used also seems to be a rather difficult one to specify properly. Of the other pumps shown in Figure 1, Pump Number 2 is a chemical feed pump and can be one of three types: diaphragm, piston (positive desplacement) or peristaltic operation. The diaphragm pump generally is used for small applications where a controlled addition is needed. The diaphragm is usually driven by a cam or a piston. The stroke is adjustable from zero to 100 percent. Capacities of this type of pump are generally in the range of 1 to 5 gph, restricting their use to chemicals that are added to maintain part per million concentrations or for pH adjustment. Materials of construction are normally plastic-polypropylene or polyethylene — with diaphragms made from any applicable type of material. Piping is simply flexible tubing and flows are so small that line loss calculations do not influence the specifications. Specifying this type of pump is relatively simple. The discharge pressure of the diaphragm pump must be greater than the line pressure. Its capacity must be adequate and the materials of construction must be compatible with the solutions to be pumped.

The piston pump is also used for chemical additions, but has a much wider range of operation than the diaphragm pump. As implied in the name, this pump is a cam driven piston. The stroke is adjustable to control the pumping rate. A wide range of sizes is available, materials of construction are metal and discharge pressures of

210

2500 psig can be obtained. This type of pump is often used for the controlled addition of large volumes of acid or caustic for neutralization processes. As with diaphragm pumps, materials of construction must be compatible with the solution being pumped. The pump capacity must be adequate and the discharge pressure must be greater than the line pressure, if the pump is discharging into a closed line. NPSH can be a problem with positive displacement pumps. It manifests itself as reduced pumping capacity instead of cavitation. Specifying a low operating speed will help eliminate this problem.

Another type of chemical addition pump is the peristaltic pump. A peristaltic pump design is normally a piece of tubing wrapped around a cam, although it can take other forms. The pump has a flooded suction, and as the cam rotates, the liquid is pushed through the tubing. Depending upon the manufacturer, some direct control of the feed rate is possible, either by changing the tubing size or by controlling the cam speed. Normally, however, the addition rate is determined by varying the solution strength. Use of these pumps is common in test systems or in temporary installations. Peristaltic pumps are the least expensive of the chemical addition pumps.

A continuing problem with this type of pump is abrasion of the tubing by the cam. It will eventually rupture; this type of pump should not be used in applications where it will be left unattended for long periods of time.

Sludge Pumps

The final type of pump used in pollution control systems is the sludge handling pump, Number 3 in Figure 1. There are two distinct types used — the progressive cavity pump and the diaphragm pump.

The progressive cavity pump is similar to a precision screw conveyor. It has two major parts — a rotor and a stator. As the rotor turns within the stator, cavities are formed which progress toward the discharge end of the pump, carrying the slurry. These pumps are self-priming, provide a uniform rate of flow and tend to be non-fouling. Particles greater than 1.5 in. in diameter can be handled, and capacities up to 1200 gpm are available. Operating pressures range from 0 to 300 psig. This type of pump is commonly used for transferring sludges with a 6 to 10 percent solids content. They have also been used for chemical slurries with a solids content of 70 percent. Operating speed depends largely on the properties of the material being handled and the pump capacity will vary for different substances. As the solids content of the fluid increases, the horsepower will increase and the operating speed will be lower. For any material, however, output per revolution will remain constant. The stator is a synthetic rubber liner, while the rotor is a chrome, tool-hardened steel. The abrasive properties of the material being tranferred will affect the pump speed. As more abrasive materials are handled, the unit capacity is reduced.

Diaphragm pumps handle a wide range of heavy slurries which are extremely difficult for other pumps to transport, with the exception of the progresive cavity pump. Although the diaphragm pump is basically a positive displacement design, its unique applications warrant discussing it separately.

The diaphragm acts as a limited-displacement piston providing pumping action when the diaphragm is forced into a reciprocating motion by a variety of means including mechanical linkage, compressed air or oil from a pulsating outside source. Connection between the energy source and the liquid being pumped is eliminated, resulting in a leak-free design. Double diaphragm pumps are also available to eliminate flow pulsations. Limitations of this design include a limited choice of materials of construction, limited head and capacity range, and the use of check valves on the suction and discharge which can become fouled. However, the diaphragm pump is an excellent, relatively low-cost unit for slurry transport.

Pumps are available in numerous design configurations: single-stage, multistage, turbine, diaphragm, inline, vertical and horizontal centrifugal, and many others. Regardless of the solution or its properties or the head required, there is a pump available that will meet the needs. The biggest problem is communicating those needs to the manufacturer in an understandable fashion to obtain the most efficient and economic pump to perform the job.

CENTRIFUGAL PUMP

The following list of operating troubles includes most of the causes of failure or reduced operating efficiency. The remedy or cure is either obvious or may be identified from the description of the cause.

Symptom A — Pump Will Not Start

Causes:

1. Blown fuses or tripped circuit breakers attributed to:
   A. Rating of fuses or circuit breakers not correct
   B. Switch (breakers) contacts corroded or shorted
   C. Terminal connections loose or broken somewhere in the circuit
   D. Automatic control mechanism not functioning properly
   E. Motor shorted or burned out
   F. Wiring hookup or service not correct
   G. Switches not set for operation
   H. Contacts of the control relays dirty and arcing
   I. Fuses or thermal units too warm
   J. Wiring short-circuited
   K. Shaft binding or sticking by reason of rubbing impeller, tight packing glands, or clogging of pump
2. Loose connection, fuse, or thermal unit

212

SELECTED MANUFACTURERS OF PUMPS FOR POLLUTION CONTROL

Manufacturer	Centrifugal Pumps	Chemical Pumps	Diaphragm Pumps	Positive Displacement Pumps	Sampling Pumps	Screw & Progressive Cavity Pumps	Submersible Pumps	Sump Pumps	Turbine Pumps	Vacuum & Air Pumps
Allis-Chalmers	●							●	●	
Atlantic Engineering Co.	●									
Armstrong Pumps Inc.	●							●		
Aurora Pump, Unit of General Signal Corp.	●							●	●	
BIF, Unit of General Signal Corp.		●	●							
Buffalo Forge Co.							●	●		
Byron Jackson Pump Div., Borg-Warner Corp.	●						●	●	●	
Calgon Corp.		●	●	●		●				
Ralph B. Carter Co.	●		●							
Celanese Piping Systems Inc.	●	●								
Chem-Tech International			●	●						
Chicago Pump Co.							●	●		
Clow Corp.	●						●	●		●
Coolant Equipment Corp.							●	●		
Crane Co.		●	●				●	●		
Crown Div., Construction Machinery Co.	●	●	●		●		●	●		
Dean Brothers Pumps Inc.	●	●					●	●		
Denver Equipment Div., Joy Manufacturing Co.	●						●	●		
Dorr-Oliver Inc.	●	●	●	●	●			●		
The Duriron Co., Inc.	●									
ECO Pump Corp.		●	●	●	●			●		
Edson Corp.		●		●						
Elders Industries, Inc.	●						●	●		
Enpo-Cornell Pump Co.							●	●		
Environment/One Corp.					●	●				
FMC Corp., Environmental Equipment Div.	●				●	●				
FMC Corp., Pump Div.	●				●		●	●	●	
Fairbanks-Morse Div., Colt Industries	●						●	●	●	
Flygt Corp.	●						●			
Fuller Co., Sutorbilt Products	●									●
The Galigher Co.	●						●	●		
Gorman-Rupp Co.	●	●	●		●		●	●		
Goulds Pumps, Inc.	●	●					●	●	●	
Harrisburg Inc.	●									
Hills-McCanna Div., Pennwalt Corp.		●	●	●						
Horizon Ecology Corp.					●					
Hydr-O-Matic Pump Co.							●	●		
ISCO		●	●	●	●					
ITT Marlow, Pump & Compression Div.	●	●	●	●	●		●	●		
Industrial Filter & Pump Manufacturing Co.	●									
Interpace Corp.		●	●							
Jaeco Pump Co.		●	●	●						
Johnston Pump Co.							●	●	●	
KSB Pump Co.	●									
Kenco Pump							●	●		
Kobe, Inc.				●	●					
Komline-Sanderson Engineering Corp.		●		●	●					

SELECTED MANUFACTURERS OF PUMPS FOR POLLUTION CONTROL

	Centrifugal Pumps	Chemical Pumps	Diaphragm Pumps	Positive Displacement Pumps	Sampling Pumps	Screw & Progressive Cavity Pumps	Submersible Pumps	Sump Pumps	Turbine Pumps	Vacuum & Air Pumps
Lakeside Equipment Corp.						●				
Liquiflo Equipment	●	●		●	●		●			
M-D Pneumatics, Inc.		●		●						●
Madden Corp.		●								
March Manufacturing Co.		●		●						
The Mec-O-Matic Co., Ecodyne Corp.		●								
Metal Bellows Corp., Zurn Industries					●					●
Mission Manufacturing Div., TRW	●	●					●			
LFE Corp., Fluids Control Div.	●	●	●	●		●	●	●	●	●
Milton Roy Co.		●								
Morris Pumps, Inc.	●							●		
Moyno Pump Div., Robbins & Meyers		●			●	●				
F. E. Myers & Bro. Co.	●	●		●			●	●		
Nagle Pumps, Inc.	●							●		
Nash Engineering Co.										●
Pacific Pumping Co.	●						●			
Pacific Pumps Div., Dresser Industries, Inc.	●									
Passavant Corp.						●				
Peabody Barnes Inc.	●						●	●		
Penberthy Div., Houdaille Industries Inc.							●	●		
Pollution Products Inc.		●								
Precision Control Products Corp.		●								
Prosser Industries Div., Purex Corp.							●		●	
Pulsafeeder/Interpace		●	●							
Roper Pump Co.						●				
Rotron Inc.										●
Science Pump Corp.						●				●
Serfilco, Div. of Service Filtration Corp.	●	●						●		
Sethco Pump Co. Inc.	●	●						●		
Sigmamotor, Inc.						●				
Singer Co., Layne & Bowler Div.		●					●	●	●	
Smith & Loveless Div., Ecodyne Corp.	●									
Summit Scientific Corp.							●	●		
Sundstrand Fluid Handling Div., Sundstrand Corp.	●									
Sykes Pumps, Inc.	●									
Tait Inc.							●	●		
Teledyne Analytical Instruments (TAI)						●				
Thomas Industries			●	●	●					●
Tuthill Pump Co.		●		●	●					
Vanton Pump & Equipment Corp.	●	●				●		●		
Vaughn Co.	●									
Wallace & Tiernan Div., Pennwalt Corp.		●	●	●						
Weil Pump Co.	●						●	●	●	
Wemco Div., Envirotech Corp.	●	●					●	●		
Wilden Pump & Engineering Co.		●	●							
A. R. Wilfley & Sons, Inc.	●									
Wilson Snyder Pump Div., U.S. Steel	●	●		●			●	●	●	
Worthington Pump Intl., Inc., Studebaker-Worthington Inc.	●	●		●			●	●	●	●

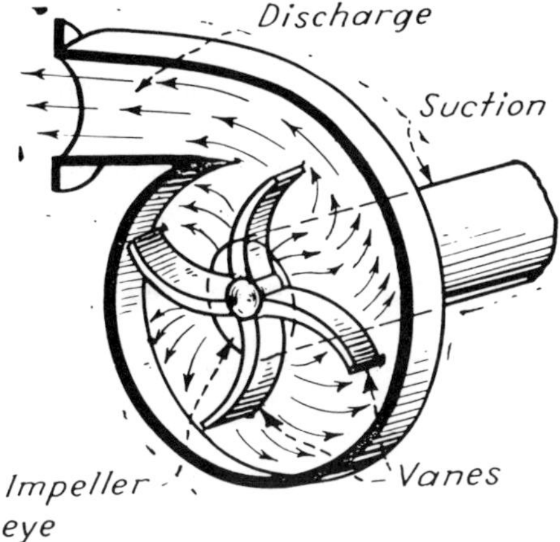

Symptom B — Reduced Rate of Discharge

Causes:

1. Pump not primed
2. Mixture of air in the wastewater
3. Speed of motor too low
4. Improper wiring
5. Defective motor
6. Discharge head too high
7. Suction lift higher than anticipated
8. Impeller clogged
9. Discharge line clogged
10. Pump rotating in wrong direction
11. Air leaks in suction line or packing box
12. Inlet to suction line too high, permitting air to enter
13. Valves partially or entirely closed
14. Check valves stuck or clogged
15. Incorrect impeller adjustment
16. Impeller damaged or worn
17. Packing worn or defective
18. Impeller turning on shaft because of broken key
19. Flexible coupling broken
20. Loss of suction during pumping may be caused by leaky suction line, ineffective water or grease seal

Sympton C — High Power Requirements

Causes:

1. Speed of rotation too high
2. Operating heads lower than rating for which pump was designed, resulting in excess pumping rates
3. Check valves open, draining long force main back into well
4. Specific gravity or viscosity of liquid pumped too high
5. Clogged pump
6. Sheaves on belt drive misaligned or maladjusted
7. Pump shaft bent
8. Rotating elements binding
9. Packing too tight
10. Wearing rings worn or binding
11. Impeller rubbing

Symptom D — Noisy Pump

Causes:

1. Pump not completely primed
2. Inlet clogged
3. Inlet not submerged
4. Pump not lubricated properly
5. Worn impellers
6. Strain on pumps caused by unsupported piping fastened to the pump
7. Foundation insecure
8. Mechanical defects in pump
9. Misalignment of motor and pump where connected by flexible shaft.

VALVES, PIPE AND FITTINGS

by James A. Edwards, Field Editor, Pollution Engineering

Valves and fittings that serve to direct and control the flow of fluids are the critical sections of a piping layout in a waste treatment process. Properly applied, they can provide either ON-OFF or proportional control, manual or automatic adjustment, good corrosion resistance, cleanouts in slurry lines, fluid blending, and backflow prevention. Good piping design depends on the proper application and use of valves and fittings through understanding of the equipment capability and how it was intended to be used.

Valves are available in a number of designs: *globe, ball, gate, plug, butterfly, diaphragm,* and *needle.* All of these types can be equipped with motor operators to automate the application. Despite the wide range of designs, all valves have only one purpose: to slow down or stop the flow of fluid. This is done by blocking the flow of the fluid with an obstruction that can be adjusted inside the pipe, with little leakage from the pipe to the outside.

Several approaches to control of fluid flow were mentioned above and each has its distinct advantages and disadvantages. Some designs provide accurate control characteristics, while others are only suitable for ON-OFF applications. In general, the better the control characteristics, the greater the pressure drop across the valve. Data illustrating this variation in pressure drop is shown in Table 1. A valve can play a major role in system pressure drop. For example, a fully open ball valve has a pressure drop equivalent to a piece of pipe the same diameter and three diameters long, while a globe valve used in the same situation has a pressure drop equivalent to a piece of pipe 340 diameters in length. Since the pressure drop is also dependent on the flow, the following example will help point out this difference between valves:

A water system is circulating 70 gpm through a 2-in.-diameter line. What increase in pressure drop will be obtained through the use of a globe valve (perpendicular stem), ball valve, and a gate valve fully open? The pressure drop is 2.87 psi/100 ft of pipe.

Using Table 1, we can develop the following chart:

Type	Equivalent Length (L/D)	Equivalent Length (ft)	Pressure Drop (psi)
Glove	340	56.6	1.63
Ball	3	0.3	0.014
Gate	13	2.16	0.062

However, if the gate valve were installed and, of necessity operated only one-quarter open, the pressure drop would increase to:

Equivalent Length (L/D)	Equivalent Length (ft)	Pressure Drop (psi)
900	150	4.33

Table 1. *Representative Equivalent Length in Pipe Diameters (L/D) of Various Valves and Fittings*

	Description of Product			Equivalent Length In Pipe Diameters (L/D)
Globe Valves	Stem Perpendicular to Run	With no obstruction in flat, bevel, or plug type seat	Fully open	340
		With wing or pin guided disc	Fully open	450
	Y-Pattern	(No obstruction in flat, bevel, or plug type seat) — With stem 60 degrees from run of pipe line	Fully open	175
		— With stem 45 degrees from run of pipe line	Fully open	145
Angle Valves		With no obstruction in flat, bevel, or plug type seat	Fully open	145
		With wing or pin guided disc	Fully open	200
Gate Valves	Wedge, Disc, Double Disc, or Plug Disc		Fully open	13
			Three-quarters open	35
			One-half open	160
			One-quarter open	900
	Pulp Stock		Fully open	17
			Three-quarters open	50
			One-half open	260
			One-quarter open	1200
Conduit Pipe Line Gate, Ball, and Plug Valves			Fully open	3**
Check Valves	Conventional Swing		0.5†...Fully open	135
	Clearway Swing		0.5†...Fully open	50
	Globe Lift or Stop;	Stem Perpendicular to Run or Y-Pattern	2.0†...Fully open	Same as Globe
	Angle Lift or Stop		2.0†...Fully open	Same as Angle
	In-Line Ball	2.5 vertical and 0.25 horizontal †...Fully open		150
Foot Valves with Strainer		With poppet lift-type disc	0.3†...Fully open	420
		With leather-hinged disc	0.4†...Fully open	75
Butterfly Valves (8-inch and larger)			Fully open	40
Cocks	Straight-Through	Rectangular plug port area equal to 100% of pipe area	Fully open	18
	Three-Way	Rectangular plug port area equal to 80% of pipe area (fully open)	Flow straight through	44
			Flow through branch	140
Fittings	90 Degree Standard Elbow			30
	45 Degree Standard Elbow			16
	90 Degree Long Radius Elbow			20
	90 Degree Street Elbow			50
	45 Degree Street Elbow			26
	Square Corner Elbow			57
	Standard Tee	With flow through run		20
		With flow through branch		60
	Close Pattern Return Bend			50

Reprinted from Crane Technical Paper No. 410.

**Exact equivalent length is equal to the length between flange faces or welding ends.

†Minimum calculated pressure drop (psi) across valve to provide sufficient flow to lift disc fully.

In short, full consideration must be given to system operating conditions and the type of control desired when selecting and sizing valves for process use.

Other factors that must be considered are the use of threaded, flanged, welded or cemented end connections, materials of construction, and packing materials.

In most waste treatment systems, the end connections will not present a problem. The systems are normally low pressure (less than 100 psi) and the end connection used will be a matter of preference or will be determined by the materials of construction.

Materials of valve construction represent a more difficult problem. They must be selected to withstand the corrosive and abrasive properties of the solution being handled. Valves are available in a wide variety of materials, ranging from the commonly used brass, steel and iron to more exotic materials such as glass. Teflon-lined units are also available. The most commonly used corrosion resistant materials are PVC and CPVC.

Manufacturers' reference tables and standard tables of corrosion data prove their value when materials are to be selected for handling corrosive solutions. It should be emphasized that a material should never be assumed resistant to corrosion by a fluid. Many problems could be avoided if available data were checked prior to system installation.

Valve packing is as important as the packing gland in a pump. It prevents costly or dangerous leaks and eliminates damage to the valve housing. Several methods are used to prevent leaks past the valve stem. Where a solution is not corrosive or hazardous, the seal is normally effected by placing packing around the valve stem and compressing it with a follower or gland, which is drawn down by a packing nut. Other common designs include the use of a bellows or a diaphragm, either of which can be made of metal or rubber.

All of the valve types mentioned above can be grouped into two fundamental types: gate and globe. This method of classification cannot be all-inclusive, since many of the designs fall in between the two extremes. However, the gate type can be considered a low pressure drop valve and the globe type a high pressure drop valve.

Gate Type Valves

This classification includes wedge, disc, double disc, plug disc, and butterfly valves. In general, these designs are all used for ON-OFF control and not for throttling of the flow, Fig. 1. This equipment is available in a wide range of sizes and is generally fabricated from bronze or steel. Problems that develop when these designs are used for throttling include an excessively high pressure drop, erosion and wire drawing of the gate and seat and the resulting failure to block the flow. Proper use of these valves includes installation as shut-off valves around equipment, bypass valves for equipment or control valves and shut-off valves on supply lines.

Globe Type Valves

This design includes *disc globe valves* (Figure 2), *angle valves* (Figure 3), *Y valves* (Figure 4), and *needle valves* (Figure 5). In general, the globe design is a high pressure drop valve which enables accurate throttling of the flow. Repeatable, stable flow settings can be obtained with valves of this type. Several factors are responsible for the performance of these valves. The high pressure drop is caused by changing the direction of flow through the valve; typically two 90-degree turns are

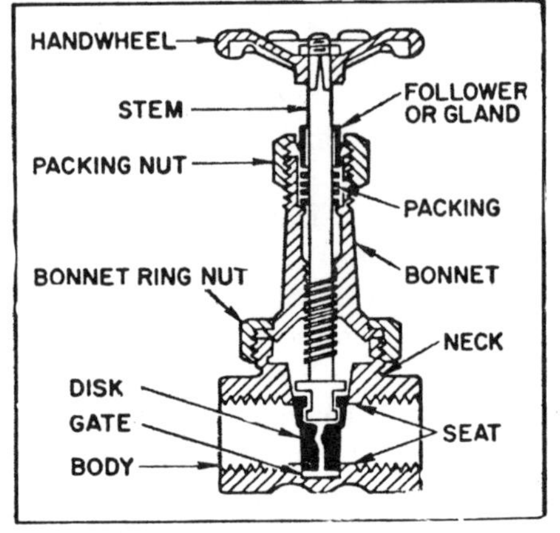

Fig. 1 Gate valve

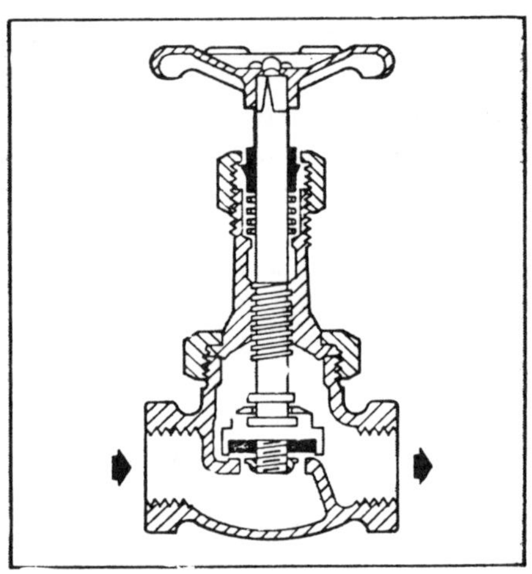

Fig. 2 Disc globe valve

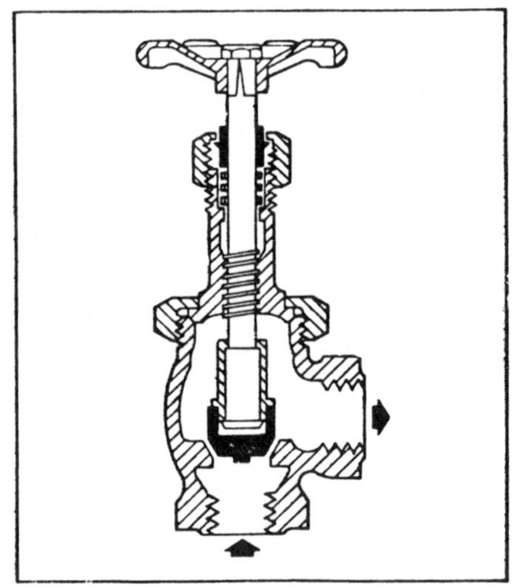

Fig. 3 Angle valve

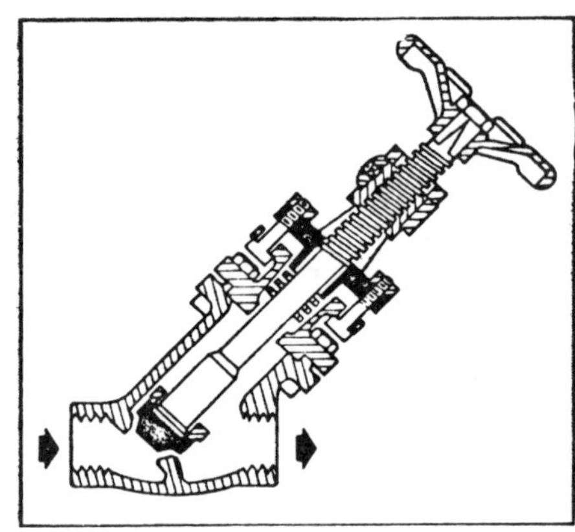

Fig. 4 Y valve

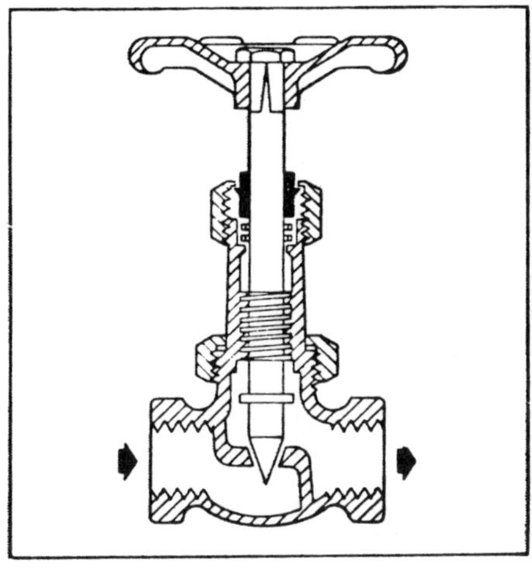

Fig. 5 Needle valve

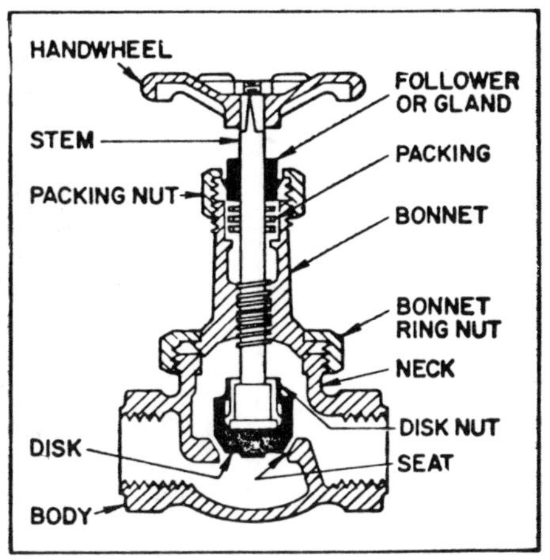

Fig. 6 Globe valve

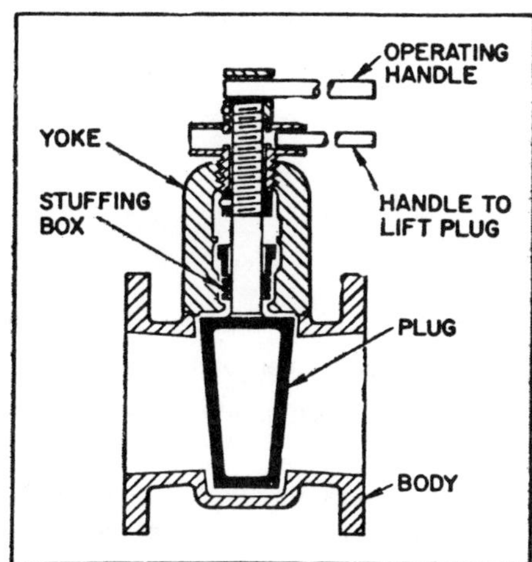

Fig. 7 Plug valve

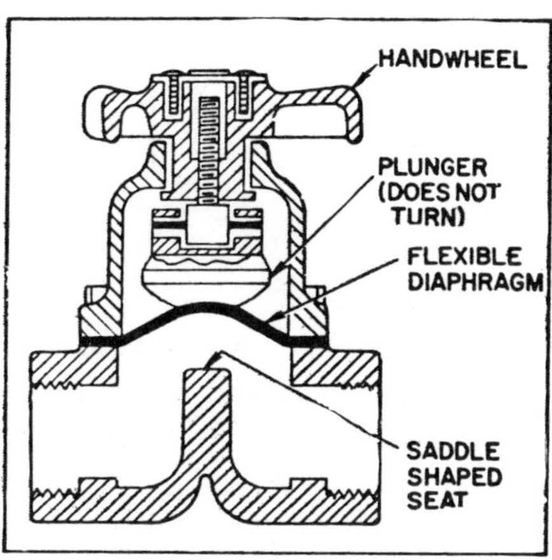

Fig. 8 Diaphragm valve

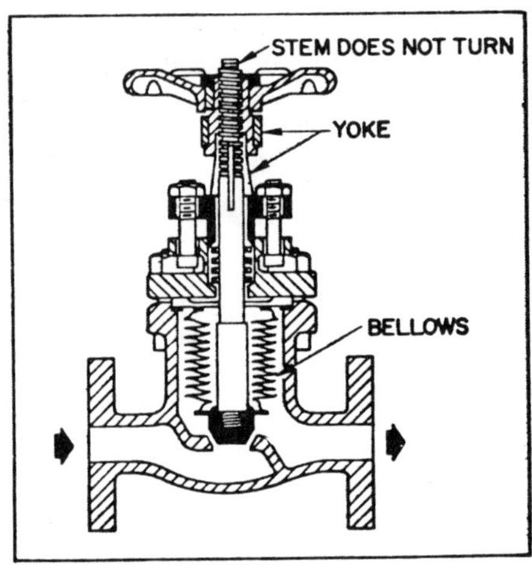

Fig. 9 Sealed bellows valve

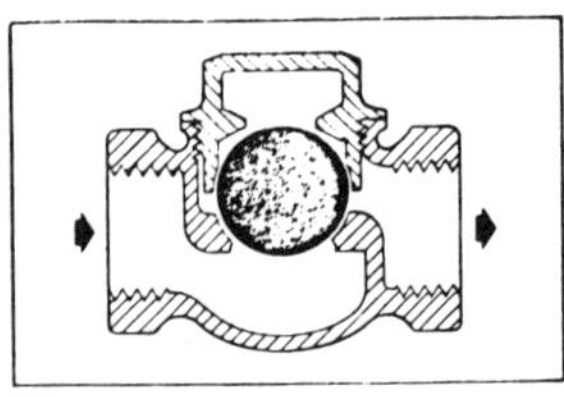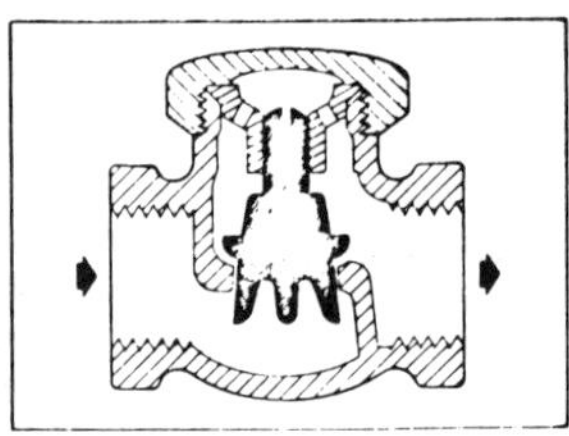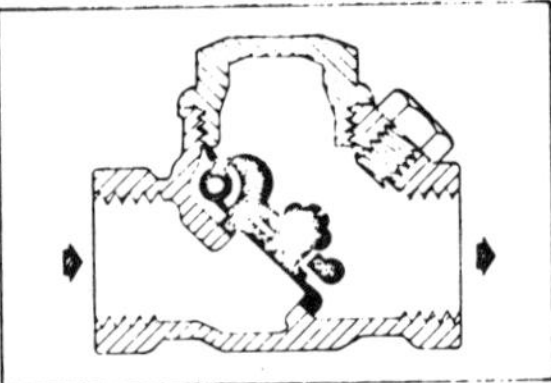

Fig. 10 Check valves

made by the fluid. As the fluid follows this flow pattern, it passes through an orifice which is partially restricted by the "globe" which can actually be a disc, needle, globe or other shape depending on the manufacturer and the desired accuracy of the flow setting.

When manual control of the flow is satisfactory, the disc design is normally used. If accurate, automatic control is required, a needle valve is most applicable — it provides sensitive adjustments proportional to flow. Seats and discs of globe type valves separate completely at the first turn of the hand wheel, eliminating the sliding friction that is a problem in gate valves. Globe valves, Figure 6, are preferred where tight seating, frequent operation, and accurate control are required. Angle valves, though similar in construction and operation to globe valves, offer two

222

added advantages. Because of the shape, two fittings can be eliminated and the pressure drop is approximately one-half that experienced with globe valves since the fluid makes only one 90-degree turn.

Ball and Plug Valves

These designs are intended for ON-OFF applications and should not be used for throttling flows. A ball or truncated cone (plug) with a hole approximately the same diameter as the pipe is used to pass or block the flow. The restricting element rotates through 90 degrees from fully open to fully closed, providing a quick shutoff when needed. Because the valve orifice is straight-through with little change in diameter from the pipe diameter, these designs provide a very low pressure drop. In some instances, the entrance and exit pressure losses may be greater than the pressure drop across the valve, Figure 7.

Diaphragm Valves

Used primarily for ON-OFF control with corrosive chemicals, diaphragm valves provide a safe way to handle or restrict the flow of compounds such as hydrochloric acid, dilute sulfuric acid, ferric chloride, and other hazardous and corrosive chemicals frequently used in waste treatment operations. The design consists of a rubber (Figure 8) or metal diaphragm (Figure 9) which is pinched against an obstruction in the valve body to block the flow. A stem-operated plunger is generally used to depress the diaphragm, although some designs use a quick-opening cam which operates through 90 degrees. Problems with this design are primarily associated with diaphragm deterioration. Since the valve head is commonly cast iron, a leak in the diaphragm often results in rapid destruction. Leaks of this type seem to occur where the valve is used only intermittently and not on a regular basis. Frequent use of the valve seems to delay cracking of the synthetic rubber diaphragms.

Check Valves

Check valves are applied to prevent an undesired flow in piping, Figure 10. Several designs are used including ball, lift, and swing checks. When the flow reverses in the line, the check valve automatically closes, preventing the backward flow of the solution. Each of these designs has distinct areas of application. Ball checks normally are used with small proportioning pumps or with diaphragm pumps on both the suction and discharge sides; swing checks are normally used in conjunction with gate valves; lift checks are used with globe valves for both liquid and vapor service.

Swing checks can be installed either horizontally or vertically. Horizontal lift checks are used on vertical lines. Swing check valves provide a lower pressure drop than lift check valves, but lift checks provide tighter seating and are used where pressure drop is not critical.

Fittings

Fittings are grouped into four distinct classifications: *branching, reducing, expanding, and deflecting.* Branching fittings include tees, crosses, side outlet elbows, and others. Reducing and expanding fittings, such as concentric reducers and bushings, change the cross-sectional area of the passageway; deflecting fittings, such as bends, elbows and return bends, change the direction of flow. In addition, other fittings such as unions and couplings are used which do not fit into these general classifications and do not offer any appreciable pressure drop. Typical pressure drop data for standard fittings is given as equivalent lengths in pipe diameters in Table 1.

The values given in this table apply only if the flow is turbulent and represents typical resistance coefficients. Detailed analysis of pressure drop in piping systems is naturally a preferred approach and should be used when designing any involved piping layout. A thorough discussion of pressure drop and piping design is available in the *Crane Technical Paper No. 410.* Additional design information can be obtained in any of the engineering handbooks or *Desing of Piping Systems,* published by M. W. Kellogg Company.

Fittings are available in almost any material desired — armored glass, stainless steel, fiberglass reinforced plastic (FRP), Teflon-lined FRP, cast iron, lined cast iron and brass. Materials are available to contain any fluid under any conditions that must be met. Several types of piping are available for waste treatment facilities, permitting the system designer some latitude in specification and material selection.

Consideration must be given to factors other than corrosion resistance and pressure drop when designing or specifying a piping system. How easy is it to install this material? Must special installation tools be rented or purchase? Can it be field fabricated? These are important considerations, for installation problems will materially increase the cost of piping systems. For example, FRP piping is lightweight and easy to handle but special flaring tools must be used to prepare the pipe for mating to flanges or fittings. Cemented joints are used and, if the workers are handling this type of pipe for the first time, faulty connections are a possibility until they become familiar with the procedure.

Rubber-lined pipe presents a different problem. It must be ordered to the exact length required, since field fabrication is difficult. Special flaring tools are required to prevent the lining from drawing into the pipe following installation. These and other problems must be considered when specifying a piping system.

Proper use of fittings is of major importance in piping design. Concentric reducers should be used whenever possible to reduce the pressure drop. The use of bushings or reducing flanges increases the line loss and will increase the energy requirements for the system. Tees can be used as mixing connections, diversion fittings, cleanout fittings or simply to mount a gage. Special fittings are available from most manufacturers and should be investigated as a need develops.

A thorough understanding of the purpose a valve or fitting is expected to serve and what it will do will result in piping designs and specifications that will meet the needs of the waste treatment system. A trouble-free system that functions as it was intended will result.

Selected Manufacturers and Suppliers

Valves

	Ball Valves	Butterfly Valves	Check Valves	Diaphragm Valves	Gate Valves	Globe Valves	Pinch Valves	Plug Valves
ACF Industries, Inc. W-K-M Valve Div.	•		•	•	•			•
Allis-Chalmers Corp.		•						
American-Darling Valve & Manufacturing Co.		•	•		•			
Armco Steel Corp., Metal Products Div.					•			
A-S-H Pump Div., Envirotech Corp.			•					
BIF Industries, A Unit of General Signal Corp.		•						
Balon Corp.	•							
Brooks Instrument Div., Emerson Electric Co.				•				
Celanese Piping Systems	•	•	•			•		
Cla-Val Co.			•	•		•		
Clow Corp., Valves Div.		•	•		•			
Crane Co.	•	•	•		•	•		
Daniel Industries			•					
Demco, Inc.	•	•			•			
DeZurik Corp., A Unit of General Signal Corp.		•	•		•			
Dow Chemical Co.			•	•				•
Dresser Mfg. Div., Dresser Industries, Inc.		•	•		•			•
Fabri-Valve		•	•		•			
Fisher Controls Co.	•	•				•	•	
Flexible Valve Corp.					•			
Garlock, Inc.		•						
Golden Anderson Valve Specialty Co.		•	•		•	•		
Hills-McCanna Div., Pennwalt Corp.	•	•			•			
Homestead Industries								•
Honeywell, Inc.	•	•				•		
ITT Grinnell Corp.					•			
Jamesbury Corp.	•	•						
Jenkins Brothers	•	•	•		•	•		
Johns-Manville Corp.			•		•			
F. B. Leopold Co., Div. Sybron Corp.		•						
Leslie Co.			•	•				
Liquid Level Lectronics, Inc.				•				
Lunkenheimer Co.	•	•	•		•	•		
MSI Industries, Inc.							•	
Mace Corp.			•	•		•		
Mueller Co.			•	•	•			•
Nupro Co.			•	•				
Peabody Doré	•							•
Henry Pratt Co.	•	•						
RKL Controls, Inc.							•	
Red Valve Co., Inc.							•	
Richards Industries, Inc., Jordan Valve Div.					•			
Robertshaw Controls Co., Industrial Instrument Div.						•		
Rockwell International	•	•	•		•	•		•
Joseph T. Ryerson & Son, Inc.						•		

Valves

	Ball Valves	Butterfly Valves	Check Valves	Diaphragm Valves	Gate Valves	Globe Valves	Pinch Valves	Plug Valves
Sethco Manufacturing Corp.	•		•	•		•		
R & G Sloane Manufacturing Co., Inc.	•		•					
Stockham Valves & Fittings	•	•	•		•	•		•
TRW Mission Manufacturing Co.		•	•					•
Taylor Instrument Process Control Div., Sybron Corp.		•		•		•		
United Brass Works, Inc.			•		•	•		
U.S. Pipe & Foundry Co.					•			
Whitey Co.	•					•		
Worcester Controls Corp.	•	•						
Zurn Industries, Hays Fluid Controls Div.	•			•				•

Pipe

	Asbestos	Cast Iron	Clay	Concrete	Ductile Iron	Fiberglass	PVC	Plastic	Polyethylene	Steel
American Cast Iron Pipe Co.		•			•					
Ameron Co.				•		•				•
Amoco Reinforced Plastics Co.								•		
Armco Steel Corp.								•		•
Can-Tex Industries			•	•			•	•		
Carlon Co.							•	•	•	
Cast Iron Pipe Research		•			•					
Celanese Piping Systems							•	•	•	
Certain-teed Products Corp.	•							•		
Ciba-Geigy Corp., Pipe Systems Dept.						•				
Clow Corp., Pipe Div.		•	•		•		•	•		
CorBan Industries, Fibreboard Corp.						•				
Dow Chemical Co.										•
E. I. du Pont de Nemours Co.									•	
Fiberex International						•				
The Flintkote Co.	•						•	•	•	
Interpace Corp.			•	•						
Johns-Manville Corp.	•						•	•	•	
F. B. Leopold Co., Div. Sybron Corp.								•		
Naylor Pipe Co.										•
Nipak, Inc.									•	
Peabody Doré								•		
Phillips Products Co., Inc.								•	•	
Joseph T. Ryerson & Son, Inc.								•	•	•
R & G Sloane Manufacturing Co., Inc.							•			
U.S. Pipe & Foundry Co.		•			•					

PREPARATION FOR SITE VISIT*

Preparation for the on-site inspection should include compilation and review of information which provides a description of the plant's physical setting, plant design details, plant operating personnel, and any available performance records, previous inspection reports, compliance orders, etc. Reference material relevant to the type of plant being evaluated should also be reviewed.

The following are specific steps suggested for use in preparation for the on-site visit:

1. Compile and Review Information on Plant's Physical Setting

Information describing the physical location of the plant will be required in the evaluation. The information should include the following:

(a) Classification of Treatment System

- Type of plant
- Contributory population (domestic, industrial, etc.)
- Wastewater system (sanitary, combined, etc.)
- Geographic-climatic effects to determine if extreme geographic-climatic areas could have an effect on plant performance. This should include both extremes of temperature and precipitation
- Determine, if possible, the characteristics of the plant's influent and effluent. This should include both values of the parameters measured by the plant (BOD_5, pH, COD, temperature, etc.), and flow quantity and variations with time.

(b) Identification of Discharge Requirements

For a particular plant location, this might consist of allowable COD, pH, BOD_5, etc., or any special controlling conditions such as a minimum DO and chlorine residuals. If the official data are not readily available from the plant, then try contacting local and/or state health departments and state, regional, and municipal pollutional control agencies. These agencies may be able to supply the needed information.

---

*Procedural Manual for Evaluating the Performance of Wastewater Treatment Plants, Contract No. 68-01-0107, USEPA, Office of Water Programs.

2. Compile and Review Information Describing the Plant and Staff

This should be accomplished in advance of the plant visit, recognizing that most state agencies will have some sort of records and previous evaluations which can be used as a comparison between the past, existing, and optimum operation of the plant. The information required to make this comparison includes:

- Size of plant (design, average daily flows, peak flows)
- Type of unit operations
- Historical operating data
- Size of staff, their qualifications, and distributions of time spent between unit operations
- Review all relevant plant documents such as design drawings, operating manuals for various pieces of major equipment, and summary (or monthly) operating reports.

This information should be compared with sections of the manual or other sources which give information on the following:

- Wastewater treatment systems operational data (use design specifications, where available)
- Personnel requirements
- Classification of wastewater treatment plants

3. Prepare a Pre-Visit Evaluation Data Sheet

Data will be needed prior to the evaluation to insure adequate data-gathering at the time of the on-site visit. The pre-visit evaluation guide should include a summary of all background information.

ON-SITE INSPECTION*

The inspection of the plant should be accomplished in several phases with each phase being progressively more detailed. The visit(s) would typically consist of two phases:

Phase One — a general orientation and overview of the plant and its operation, including:

- Meet initially with plant engineer or chief operator. Have him describe the plant and its principal operating characteristics, on a schematic basis (this and the following steps are to help orient you to the plant, but also to give you an indication of how well the staff understands the system).
- Interview plant staff members, starting with plant superintendent and

---

*Always contact the plant manager in advance to set up an appointment; avoid ill-will from arriving unannounced.

other supervisory personnel and working progressively through the appropriate operating personnel.

- Determine routine plant performance and compare this with design performance and norms section of the manual.

Phase Two — problem identification with an evaluation as to effect on overall plant performance. This phase also includes the procedure for laboratory evaluation.

Problem identification begins with a tour of the facility. Have the plant engineer or operator give you a complete tour of the facilities. Watch for and inquire about:

- Excessive particles and/or floc flowing over overflow weirs
- Excessive grease and scum buildup
- Any unusual equipment such as special pumps, chemical feeders, temporary construction on structures or other jury-rigged systems which are being used to correct problems (or possibly causing them)
- Evidence of flow in by-pass channel to parallel units because problems have come up in normal operating units
- Excessive odors
- Abnormal color of wastewater in various process stages.

If any special plant modifications were made, determine:

- Purpose
- Physical makeup
- Effect on the other treatment processes by comparing old operational data with data after modification.

REVIEW RECORDS

In general, all records should be analyzed and compared with the information in the manual and/or other sources for consistency, method of calculation and to verify that recorded values are within the range recommended by this manual or other sources.

The following records should be reviewed:

FLOW Hydraulic data are reviewed for:

- Consistency with the design flow and with present population served
- Over- and under-loading of the various treatment units.
- Meter calibration

UNIT OPERATIONAL DATA (BOD, COD, Suspended Solids, etc.) are reviewed for:

- Consistency with design specification and values indicated in the manual or other sources
- Extreme values for the daily flows.

POWER CONSUMPTION records should be reviewed for values above or below normal. These records would tend to indicate the following:

- Operating heads lower than the pump's rating
- Specific gravity or viscosity of liquids being pumped is too high.

POWER CONSUMPTION AND FLOW RECORDS. Analyses of the FLOW data in combination with POWER might indicate the following:

- Output of each pump separately and pumps collectively
- Unusual operating conditions which are in effect or have occurred
- Changes in efficiency of pumps by comparison of gallons pumped/kilowatt hour over an extended period of time.

MAINTENANCE DATA*. The maintenance program is usually a good indicator of operational quality; this can be indicated by checking:

- Manufacturer's maintenance schedule for components
- Type of routine being used for maintenance scheduling
- Personnel qualifications for the type of maintenance work being performed.

Procedures for Problem Evaluation

In general, the problems detailed in the manual are those most commonly encountered. However, these procedures can be used for any type of problem evaluation. The first step in problem evaluation is to determine if the plant is meeting design performance standards by comparing its effluent quality and overall removal efficiencies with those specified by the design. If the plant does not routinely meet performance specifications, it will be necessary to determine whether the deficiency is due to problems which fall into two categories:

PROBLEM DEFINED — If the treatment plant operator has defined the problem:

a.   Verify general area of problems, such as related to process, maintenance or

---

*Refer to O & M manual requirements.

design, sampling, etc.

b. For common process problems, refer to that section of the manual dealing with the problem.

c. Develop sampling and testing program to provide additional data, if needed.

PROBLEM UNSPECIFIED — If effluent discharge does not meet required standards and no definite problem area has been established:

a. Review flow and process records again in greater detail.

b. Recheck sampling and testing procedures required.

c. Compare sampling and testing program against recommended programs in the manual.

d. Recommend a modified testing and sampling program to furnish additional data for evaluation.

e. Compare the data with the problem indicators detailed in the manual to see if there is a solution offered.

f. For those problems not specifically covered in the manual, and if the evaluator's experience does not suffice, it should be recommended that a consultant be hired.

MAINTENANCE PROBLEM — Refer to sample maintenance program and compare with actual plant program; recommend new program where needed.

Total Plant Evaluation

This should include the following:

1. TOTAL EVALUATION OF PLANT — utilizing the Evaluation Guide materials at the end of this section

   (a) Modification of initial evaluation, if appropriate
   (b) Differences in existing plant performance and operational data with design and/or manual operational or performance data
   (c) Personnel needed for adequate operation[*]
   (d) Type of sampling program required to give needed data
   (e) Maintenance system needed
   (f) Laboratory equipment needed
   (g) Problems encountered
   - Those corrected by visit
   - Those that need outside help to correct
   - Proposed solutions.

---

*See EPA Staffing Guides.

2. **FINAL REPORT** — should contain the following elements:

    (a) Summary of on-site visit

    (b) A list of problems encountered

    (c) Solutions recommended

    (d) Proposed action.

<u>Pre-Visit Evaluation Guide</u>

Plant Identification (Name, Owner, etc.) _______________________

Plant Location____

Operator in Charge____

Date of Evaluation____

Evaluation by____

Date of Plant Construction ____________________________________

Name of Design Firm ____

Regulatory Agency of Concern___________________________________

Stage Operating Permit? _______________________Yes__ No__ Number______

Background Information (obtain prior to visit whenever possible)

    1. Type of Plant_____________________________

    2. Schematic of Flow Route to Units on Line

    3. Contributory Population

          Domestic _______________________

          Industrial (P.E.) _______________

          Other _________________________

    4. Type of Wastewater System

          Combined _______________________

          Sanitary_______________________

          Industrial_____________________

    5. Geographic/Climatic Effects

          Temperature Ranges (^{o}F) _______ to _______

          Rainfall Extremes (in.) _______ to _______

6. Plant Wastewater Characteristics

	Nitro-gen (mg/ℓ)	Total Phos-phorus (mg/ℓ)	BOD (mg/ℓ)	Sus-pended Solids (mg/ℓ)	COD (mg/ℓ)	Flow (MGD)	Dis-solved Oxygen (mg/ℓ)
a. Plant influent							
b. Plant effluent							
c. Overall perform-ance (%)							
d. Design and/or manual recommended performance values (%)							
e. Existing receiving water quality							
f. Required receiving water quality							

7. Possible Problems

    a. Identified from operating reports

    b. Identified from previous inspection reports

    c. Complaints

<u>On-Site Evaluation Guide</u>

1. Flow

    Design (actual)____

    Daily Average____

    Peak ____

    Note any variation or erratic flow patterns.

2. Process Units Employed and All Pertinent Information

Unit Process	Operational Parameters		Loading Rates	
	Existing Plant Values	Design and/or Recommended Manual Values	Existing Plant Values	Design and/or Recommended Manual Values

3. Historical Operational Data

    ● Organize data to see if following occurred:

      Equipment failure - when, what type failure, how long out
      of service

      Extreme weather conditions

    ● Excessive loadings on plant:

      Flow -- when, how long, results of

      Organic - when, how long, results of

- Changes in process operation, such as:

    Increasing air flow to activated sludge units

    Decreases in detention times

    What caused changes? Are changes still in effect? Why?

4. Plant Personnel

Size of Staff for Daily Average Flow Handled		Qualifications		Types of Shifts (hours)	
Existing Plant	Recommended Manual and/or Other Sources	Existing Personnel	Recommended Manual and/or Other Sources	Existing	Recommended Manual

5. Laboratory Evaluation

Type of Tests Performed for Treatment System Evaluated			Testing Procedures and Equipment Used	Type of Tests Performed for Treatment System as per Manual
Type	Frequency	Location		

	Yes	No
I. OPERATING PROBLEMS • Are there problems affecting the performance of this plant? • Can they be solved without major construction? • Are the skills to solve the problem available among the staff? • Was a solution suggested by the evaluator? • Is it a permanent solution?		
II. THE SAMPLING AND TESTING PROGRAM • Are the sampling locations suitable? • Are testing procedures "Standard methods?" • Is testing frequency adequate to maintain process control?		
III. LABORATORY FACILITIES • Is the laboratory well-organized and apparently functioning as intended? • Is there enough equipment to perform all the necessary tests? • Is the equipment being properly used?		
IV. PERSONNEL – Plant Operators (including lab personnel) • Is the staff adequate in size? • Are they qualified? State certified? • Is there an operator training program? • Are the shifts adequate and balanced?		

(continued)

V. OVERALL PLANT PERFORMANCE

- Is it operating to its design specifications?

- Is it meeting discharge requirements for its location?

- Even if adequate, can plant performance be improved by simple and/or inexpensive changes?

- _________________________

Evaluator: If any of the answers are no, the following steps should be taken:

1. Submit a report which includes:

    - Summary of on-site visit

    - A list of problems encountered

    - Solutions recommended

    - Proposed action.

2. Discuss the recommendations with treatment plant staff.

3. In cooperation with the plant operator (and local officials, if necessary), decide on a course of action to solve the problem(s).

4. After a suitable period of time, revisit the plant to make a re-evaluation.

APPENDIX F

SEWER SYSTEM EVALUATION*

1:0 INTRODUCTION

Extraneous water from infiltration/inflow sources reduces the capability of sewer systems and treatment facilities to transport and treat domestic and industrial wastewaters. Infiltration occurs when existing sewer lines undergo material and joint degradation and deterioration as well as when new sewer lines are poorly designed and constructed. Inflow normally occurs when rainfall enters the sewer system through direct connections such as roof leaders and catch basins. The elimination of infiltration/inflow by sewer system rehabilitation can often substantially reduce the cost of wastewater collection and treatment. However, a logical and systematic evaluation of the sewer system is necessary to determine the cost-effectiveness of any sewer system rehabilitation to eliminate infiltration/inflow.

The Federal Water Pollution Control Act Amendments of 1972 require that after July 1, 1973, all applicants for treatment works grants must demonstrate that each sewer system discharging into the treatment works is not subject to excessive infiltration/inflow. The requirement was implemented in the Rules and Regulations for Sewer System Evaluation and Rehabilitation, 40 CFR 35.927.

This document is intended to provide engineers, municipalities, and regulatory agencies with guidance on sewer system evaluation.

2.0 INFILTRATION/INFLOW ANALYSIS

The infiltration/inflow analysis should provide the information necessary to establish the non-existence or possible existence of excessive infiltration/inflow in the sewer system(s) and justify any proposed sewer system evaluation survey.

The analysis should include each sewer system tributary to the treatment works project. The sewer system(s) should not be limited by political jurisdictions or sewer types. The treatment works grant applicant is responsible for the entire sewer system evaluation and any rehabilitation to eliminate excessive infiltration/inflow. The sewer system included in the evaluation should originate at the wastewater sources, such as commercial buildings or private residences, and terminate at the wastewater facility.

The estimated flow rates of infiltration/inflow, infiltration and inflow entering the sewer system should be stated in the analysis. The following diagram identifies these terms:

---

*Guidance for Sewer System Evaluation, USEPA, Office of Water Program Operations, March 1974.

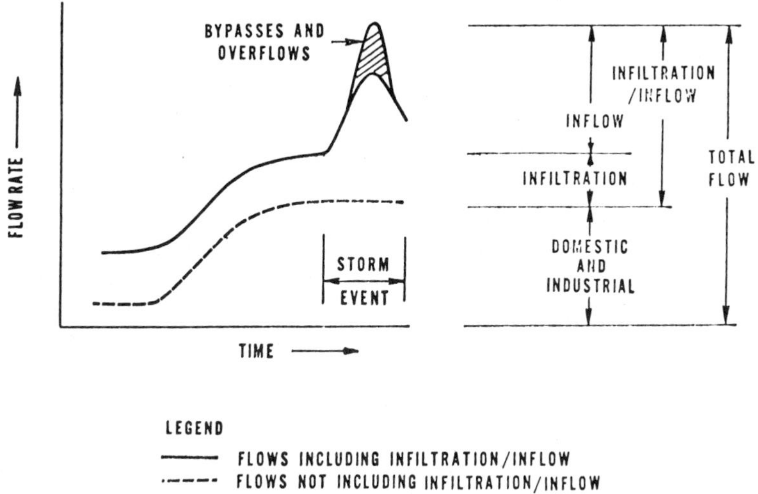

The difference between the maximum domestic and industrial flow rate and total flow rate would represent the total infiltration/inflow entering the sewer system. The difference between the maximum domestic and industrial flow rate and the maximum flow rate during periods of high ground water (with no rainfall) normally represents the infiltration entering the sewer system. The amount of flow increase during storm events (including bypasses and overflows) normally represents the inflow entering the sewer system.

Data sources for the analysis should include maps, operation and maintenance records, observations by past and present municipal employees, and previous engineering reports. When complete flow records are not available, estimated flow rates may be computed from observed flow depths. Data presented in the analysis does not have to be based on absolute measurements. A physical examination of key manholes is normally conducted to obtain data for the analysis.

Estimated flow data should be related to rainfall intensity or other pertinent data. A rainfall and sewage flow hydrograph should be included in the analysis. Each bypassed flow and when possible overflows should be identified by location, cause, duration, quantity, frequency, rate and method of discharge from the system.

The total domestic and industrial wastewater flow rates and their relationship to water consumption plus the domestic wastewater flow per capita should be stated in the analysis.

A general description of the geographical and geological characteristics of the area served by the sewer system should be presented in the analysis. This description should include soil types, topography, rainfall data, known ground water levels and other pertinent information.

The general discussion of a sewer system in the analysis should include: the type of sewer system, i.e. sanitary or combined sewers; the known methods of sewer construction; the maximum, minimum, and average depth of the sewers; major known sources of inflow; the structural condition, operation and maintenance practices, amount and type of deposits, degree of root intrusion, and other pertinent sewer system information; plus an evaluation of the probability of future decreases or increases in the quantities of infiltration/inflow.

A comparison of the cost estimates for transportation and treatment of the infiltration/inflow versus correction of the infiltration/inflow is normally sufficient to determine if infiltration/inflow is non-excessive or possibly excessive. Treatment costs should be based on achieving the effluent limitations that are or will be included in the NPDES permit(s) for discharges from the system.

When a sewer system has bypasses or overflows due to combined sewers and there is or will be no control or treatment required of the bypasses or overflows in the NPDES permit, treatment costs should be based on treatment of the total flow minus the bypasses or overflows attributable to the combined sewer inflow. In those cases where control or treatment of combined sewer bypasses or overflows is required, the cost-effectiveness analysis should be based on control or treatment of the total flow in the system. In all instances, the excessive infiltration should be eliminated from the entire sewer system including the combined sewer portions.

Infiltration/inflow correction cost estimates should include the costs of an evaluation survey, sewer system rehabilitation, and transportation and treatment of the infiltration/inflow not eliminated by rehabilitation.

3.0 SEWER SYSTEM EVALUATION SURVEY

The sewer system evaluation survey is a systematic examination of the sewer system to determine the specific location, flow rate and rehabilitation costs of the infiltration/inflow problem. The following approach is designed to avoid overstudy of the infiltration/inflow problem, including unnecessary sewer cleaning and internal inspection. Each phase of the evaluation survey is supported by the preceding phase.

The evaluation survey is normally divided into five consecutive phases: (1) physical survey, (2) rainfall simulation, (3) preparatory cleaning, (4) internal inspection, and (5) survey report. However, in certain situations, it will be possible to acquire the desired information and results more economically be combining or eliminating certain phases of the survey. The physical survey and rainfall simulation phases may provide sufficient data to determine the existence or non-existence of excessive infiltration/inflow. In such cases, the cleaning and internal inspection phases could be eliminated.

3.1 Physical Survey

The first phase of the sewer system evaluation survey should be a physical survey to determine the flow characteristics, ground water levels and physical conditions of the sewer system.

In the first step of the physical survey, flow characteristics, and, if infiltration is a problem, ground water levels at key manholes in the sewer system are determined. Evaluation of this data would enable identification of segments of the sewer system requiring further study. In certain instances, the study area for the sewer system can be determined from data acquired during the infiltration/inflow analysis.

The second step of the physical survey should be an examination of each manhole in the study area to determine the actual physical condition of the sewer system. This examination involves a physical lamping of each pipeline connected to the manholes. This data should aid in the identification of infiltration/inflow sources and provide a factual base for any sewer cleaning.

3.2 Rainfall Simulation

The second phase of the evaluation survey should be rainfall simulations to identify sections of sewer lines which have infiltration/inflow conditions during periods of rainfall.

Dyed water flooding of storm sewer sections which parallel or cross sanitary sewer sections (including service connections) and have crown elevations greater than the invert elevations of the sanitary sewers is a method of conducting the rainfall simulation phase. Stream sections, ditch sections, and ponding areas located near or above sanitary sewer sections should be dyed water flooded to identify other sources of infiltration/inflow. The downstream sanitary manhole is monitored for evidence of dyed water. The observed presence, concentration, and travel time of the dyed water into the sanitary sewer can be correlated with the soil types to obtain an estimate of the sources and quantities of infiltration/inflow. If the sewer system does not contain water traps or sagged lines, smoke testing could be used to identify connections from catch basins, roof leaders, yard drains and area drains.

3.3 Preparatory Cleaning

The third phase of the evaluation survey should be the preparatory cleaning of selected sewer lines to provide for unobstructed internal inspection. The selection of sewer sections for internal inspection is determined by analysis of the data from the physical survey and rainfall simulation phases. Selected sewer sections should have obvious potential for excessive infiltration/inflow and warrant the necessary preparatory cleaning and internal inspection.

3.4 Internal Inspection

The fourth phase of the evaluation survey should be the internal inspection of selected sections of the sewer system. This phase should determine the specific location, condition, estimated flow rate, and cost of rehabilitation for each source of infiltration/inflow defined in the selected sections. A descriptive record of all structural defects, service connections, abnormal conditions and other pertinent observations should be obtained during the inspection. The source of service connection flows should be identified. An estimated flow rate is determined for each infiltration/inflow source.

Internal inspection for infiltration conditions is normally conducted during periods of maximum ground water levels. One exception to this procedure is when the sewer is located above the maximum ground water level. All storm sewers sections, stream sections, ditch sections, and ponding areas related to the infiltration/inflow conditions are normally flooded during the internal inspection.

The method used for internal inspection of sewer sections should be the best and most cost-effective method of obtaining the necessary information. Television is an acceptable method of obtaining the necessary information. Inspection of large sewers may be accomplished by actual observation. Photographs or video tapes of infiltration/inflow sources can be used to support the field data.

3.5 Survey Report

The final phase of the evaluation survey should be a survey report of the data gathered during the survey, plus a justification for each sewer section cleaned and internally inspected, (costs not justified will be unallowable grant costs) and a proposed rehabilitation program to eliminate all defined excessive infiltration/inflow.

Each source of infiltration/inflow found during the survey should be identified in the report by specific location, condition, flow rate, method and cost of rehabilitation, and cost of transportation and treatment. An infiltration/inflow source should be proposed for rehabilitation if the rehabilitation cost does not exceed the cost of transportation and treatment.

Rehabilitation costs for an infiltration/inflow source should be based on the most cost-effective method of rehabilitation. (Several sources in a sewer segment between two consecutive manholes could be combined to achieve this objective.) Methods of rehabilitation can include: (1) replacement of sewer sections or service connections; (2) insertion of sewer liners; (3) internal or external pressure grouting with chemical sealants; (4) removal or plugging of inflow connections; (5) manhole grouting; and (6) replacement, elevating and/or sealing of manhole covers. Cement mortar grouting is not an effective method of rehabilitation except for manholes. Chemical sealants used for pressure grouting should have the demonstrated capability to eliminate infiltration under similar soil and sewer conditions. When pressure grouting is the selected method of rehabilitation, the estimated cost for the chemical sealant must be included in the rehabilitation costs.

242

When the sewer system contains a portion of combined sewers, the major sources of inflow in the sanitary sewer protions tributary to the combined sewer portions, such as cross connections from storm sewers, yard and area drains, roof leaders, manhole covers and catch basins should be proposed for rehabilitation. If control or treatment is or will be required for bypasses or overflows in the NPDES permit, the remaining inflow sources in the sewer system should be proposed for rehabilitation if the cost of rehabilitation does not exceed the cost of transportation and treatment.

Basic Laboratory Procedures

1. Effluent Monitoring Procedure: Determination of Five-day Biochemical Oxygen Demand (BOD_5)
2. Effluent Monitoring Procedure: Winkler Determination of Dissolved Oxygen-Azide Modification
3. Effluent Monitoring Procedure: Determination of Dissolved Oxygen Using A Dissolved Oxygen Meter
4. Effluent Monitoring Procedure: Measurement of pH
5. Effluent Monitoring Procedure: Total Suspended (Non-Filterable) Solids, mg/liter
6. Effluent Monitoring Procedure: Reporting of Self-Monitoring Data

REFERENCES

Self-Monitoring Procedures: Basic Parameters for Municipal Effluents Student Reference Manual, EPA-430/1-74-015, USEPA, Water Programs Operations.

EFFLUENT MONITORING PROCEDURE: DETERMINATION OF FIVE-DAY BIOCHEMICAL OXYGEN DEMAND (BOD$_5$)

1. Analysis Objectives:

The learner will determine the five-day biochemical oxygen demand of a sewage sample.

2. Brief Description of Analysis:

The sample is diluted with a high quality distilled water containing nutrients salts and a buffer. Two biochemical oxygen demand (BOD) bottles are filled with the diluted sample. The dissolved oxygen (DO) content of the first bottle is determined, and expressed as mg of DO/liter. The second bottle is stored in the dark at 20°C for five days. During the five-day period, microorganisms in the sample break down complex organic matter in the sample, using up oxygen in the process. At the end of the five-day period, the DO content of the second BOD bottle is determined, and again expressed as mg of DO/liter. The depletion in oxygen content, divided by the percent of sample used (expressed as a decimal fraction) is the five-day biochemical oxygen demand expressed as milligrams of BOD per liter of sample, BOD$_5$ is the symbol for the five-day biochemical oxygen demand.

General Description of Equipment Used in the Process

A. Capital

1.   Trip balance, 100 g. capacity
2.   Still, or other source of distilled water
3.   Incubator capable of maintaining a temperature of 20°C + 1°C, and large enough to hold four 300 ml BOD bottles and a $\overline{3}$ liter jug or bottle

B. Reusable

1.   Brushes (for cleaning glassware)
2.   Brush (for cleaning balance)
3.   Laboratory apron
4.   Safety glasses
5.   One spatula (medium size)
6.   One distilled water plastic squeeze bottle
7.   One pen or pencil
8.   One notebook (for recording data)
9.   Seven plastic weighing boats (2–3 inches square)

10. Sponges (for cleaning of laboratory table tops)
11. One 3 liter jug or bottle with narrow neck
12. One powder funnel about 3 inch diameter
13. One 1 liter volumetric flask
14. Four 1 liter glass stoppered bottles
15. Two 1 liter graduated cylinders
16. One siphon (long enough for use with the 1 liter graduated cylinder)
17. Four 1 ml volumetric pipets
18. One 10 ml volumetric pipet
19. One 20 ml volumetric pipet
20. One plunger type mixer (for use with the 1 liter graduated cylinder)
21. Four 300 ml BOD bottles
22. Equipment for doing a Winkler DO determination-azide modification, see EMP CH.OEMP. 1.8.74, Winkler Determination of Dissolved Oxygen-Azide Modification, or
23. One dissolved oxygen meter, see EMP CH.Odo.EMP.1.8.74, Determination of dissolved Oxygen Using a Dissolved Oxygen Meter
24. One 2 liter beaker (for preparing cleaning solution)
25. One 12 inch stirring rod (for preparing cleaning solution)

Consumable

1. Small wad of cotton (to plug the 3 liter jug or bottle)
2. 8.5 g. of potassium dihydrogen phosphate, KG_2PO_4
3. 21.75 g. of dipotassium hydrogen phosphate, K_2HPO_4
4. 33.4 g. of disodium hydrogen phosphate heptahydrate, $Na_2HPO_4.7H_2O$
5. 1.7 g. of ammonium chloride, NH_4Cl
6. 22.5 g. of magnesium sulfate heptahydrate, $MgSO_4.7H_2O$
7. 27.5 g. of anhydrous calcium chloride, $CaCl_2$
8. 0.25 g. of ferric chloride hexahydrage, $FeCl_3.6H_2O$
9. Reagents for doing a Winkler DO determination-azide modification, see EMP CH.OEMP.la.9.74, Winkler Determination of Dissolved Oxygen-Azide Modification
10. Reagents for use with a dissolved oxygen meter, see EMP CH.O.do.EMP .la.9.74, Determination of Dissolved Oxygen Using a Dissolved Oxygen Meter
11. Concentrated sulfuric acid, H_2SO_4
12. Potassium dichromate, $K_2Cr_2O_7$
13. Soap

(Items 11, 12, and 13 are for cleaning glassware. The quantities needed will therefore vary).

All reagents should be of high quality. Different chemical manufacturers may have different ways of indicating a high quality reagent. While no endorsement of one chemical manufacturer over another is intended, the following are some designations used in four chemical catalogs to indicate high quality reagents.

Catalog	Designations
Thomas	Reagent, ACS, Chemically Pure (CP)
Matheson, Coleman & Bell	Reagent, ACS
Curtin Matheson Scientific, Inc.	Primary Standard, ACS, AR
Fisher	Certified, ACS

EFFLUENT MONITORING PROCEDURE: Determination of Five-Day Biochemical Oxygen Demand(BOD_5)

OPERATING PROCEDURES	STEP SEQUENCE	INFORMATION/OPERATING GOALS/SPECIFICATIONS	TRAINING GUIDE NOTES
A. Equipment Preparation 1. Cleaning of glassware	1. Clean all glassware and rinse with distilled water.		V.A.1.1
2. Balance inspection	1. Check all balances for cleanliness and proper operation.		
B. Reagent Preparation 1. Distilled water	1. Distill 2 liters of water into a small neck jug (or large bottle).	1a. Unless otherwise specified, the term water means distilled water. 1b. Unless otherwise specified, solutions should be stored in glass stoppered bottles.	
	2. Plug the jug with a loose fitting piece of cotton.	2a. Air should be able to pass freely into the jug.	
	3. Store the jug at 20°C $\pm$ 1°C for 48 hours prior to use,	3a. This length of time has been determined simply on the basis of experience.	
	4. or aerate the water just prior to use.	4a. Do this by shaking the water in a half-filled jug, 4b. or by using a clean supply of compressed air. (Be cautious about air jets and motors which may simply contaminate the water with oil).	
2. Phosphate buffer solution	1. Weigh 8.5 g. of potassium dihydrogen phosphate, KH_2PO_4.		
	2. Weigh 21.75 g. of dipotassium hydrogen phosphate, K_2HPO_4.		
	3. Weigh 33.4 g. of disodium hydrogen phosphate hepta-hydrate, $Na_2HPO_4 \cdot 7H_2O$		

EFFLUENT MONITORING PROCEDURE: Determination of Five-Day Biochemical Oxygen Demand (BOD_5)

OPERATING PROCEDURES	STEP SEQUENCE	INFORMATION/OPERATING GOALS/SPECIFICATIONS	TRAINING GUIDE NOTES
2. Continued	4. Weigh 1.7 g. of amm[on]. chloride, NH_4Cl.		
	5. Dissolve the four c[hemi].cals together in about 5[00] ml of water.		
	6. Dilute to 1 liter.		
3. Magnesium sulfate solution	1. Dissolve 22.5 g. of magnesium sulfate heptahydrate, .$MgSO_4 \cdot 7H_2O$, in water and dilute to 1 liter.		
4. Calcium chloride solution	1. Dissolve 27.5 g. of anhydrous calcium chloride, $CaCl_2$, in water and dilute to 1 liter.		
5. Ferric chloride solution	1. Dissolve 0.25 g. of ferric chloride, $FeCl_3$, in water and dilute to 1 liter.		
C. Procedure 1. Sample dilution	1. Siphon about 500-ml of the 20°C water into a 1 liter graduated cylinder.	1a. Do not cause splashing which might change the oxygen content of the water.	
	2. Add 1.0 ml of each of the magnesium sulfate, calcium chloride, ferric chloride, & phosphate buffer solutions.	2a. One ml of each of the 4 solutions is used for each liter of water. 2b. These 4 solutions may be added to the reservoir of 20°C water just prior to use.	
	3. Shake the sample container, measure the sample, and add it to the graduated cylinder.	3a. For influents of domestic wastewaters: 10.0 ml (1% of the liter volume) 20.0 ml (2% of the liter volume) 40.0 ml (4% of the liter volume)	

EFFLUENT MONITORING PROCEDURE: Determination of Five-Day Biochemical Oxygen Demand (BOD_5)

OPERATING PROCEDURES	STEP SEQUENCE	INFORMATION/OPERATING GOALS/SPECIFICATIONS	TRAINING GUIDE NOTES
1. Continued		For effluents from primary treatment plants: 40.0 ml (4% of the liter volume) 60 ml (6% of the liter volume) 80 ml (8% of the liter volume) For effluents from secondary treatment plants: 200 ml (20% of the liter volume) 300 ml (30% of the liter volume) 400 ml (40% of the liter volume) 3b. The sample volumes above are suggested values. The actual sample volume for each kind of waste must be determined by experience.	
	4. Siphon in additional 20°C water to the 1 liter mark.	4a. Do not cause splashing. 4b. Other sample dilution methods are sometimes used; e.g., a 1 liter volumetric flask in place of the graduated cylinder.	
	5. Use a plunger-type mixer to mix the contents of the cylinder.	5a. Mix gently so as not to cause splashing.	
2. BOD bottle filling	1. For each sample volume used, fill 2 BOD bottles by siphoning from the liter cylinder.	1a. Hold the end of the siphon near the bottom of the BOD bottle so as to prevent splashing. 1b. Open the siphon slowly.	
	2. Stopper the BOD bottles.	2a. Do not cause formation of an air bubble by inserting the stopper too vigorously.	
3. Blank determination	1. Siphon about 500-ml of the 20°C water into a 2nd one liter graduated cylinder.	1a. Cause no splashing.	
	2. Add 1.0 ml each of the magnesium sulfate, calcium chloride, ferric chloride, and buffer solutions.		

EFFLUENT MONITORING PROCEDURE: Determination of Five-Day Biochemical Oxygen Demand (BOD_5)

OPERATING PROCEDURES	STEP SEQUENCE	INFORMATION/OPERATING GOALS/SPECIFICATIONS	TRAINING GUIDE NOTES
3. Continued	3. Siphon in additional 20°C water to the 1 liter mark.	3a. Cause no splashing. 3b. No sample is used in this second cylinder.	
	4. Mix the contents of this second cylinder using a plunger-type mixer.	4a. Mix gently.	
	5. By siphoning, fill 2 BOD bottles with this mixture.	5a. These two bottles are called blanks.	
4. DO determination	1. Fill the flared top of one of the sample and one of the blank BOD bottles with water.		
	2. Store them at 20°C in the dark for 5 days.	2a. Check the flared tops daily and refill with water if necessary. 2b. These are alternate ways of storing the bottles at 20°C so as to maintain a water seal.	
	3. Determine the DO of the 2nd sample and blank bottles.	3a. Use the Winkler method-azide modification, or a dissolved oxygen meter. 3b. This determination should be done within 15 minutes after filling the BOD bottles. 3c. These DO values are often called initial DO values.	
	4. After 5 days, determine the oxygen content of the stored sample and blank BOD bottles.	4a. Use the same method as before.	
5. Calculations	1. Subtract the DO (expressed in mg/l) value of the fifth day sample bottle from the initial DO (expressed in mg/l) value of the first bottle.	1a. e.g. 5.0 = mg initial DO in the first bottle, and 2.0 = mg DO in the stored bottle after 5 days. 1b. 5.0 - 2.0 = 3.0	

OPERATING PROCEDURES	STEP SEQUENCE	INFORMATION/OPERATING GOALS/SPECIFICATIONS	TRAINING GUIDE NOTES
5. Continued	2. Divide the difference by the percent of sample used, expressed as a decimal; the answer is mg BOD_5/l.	2a. e.g., if 10% sample was used, then mg BOD_5/liter = 3.0/0.1 = 30. 2b. If the decrease in DO over the 5 day period is not at least 2.0 mg/liter, repeat the test using more sample. 2c. If there is not at least 1 mg of DO left in the stored bottle after 5 days, repeat the test using less sample. 2d. It is common to set up at least 3 dilutions of a particular sample so that 1 of the bottles will show an acceptable oxygen depletion over the 5 day period.	
	3. For the blank BOD bottles, subtract the ml of sodium thiosulfate titrant used for the stored bottle from the ml used for the initial bottle.	3a. The difference should not be greater than 0.2 ml. 3b. If it is, the 20°C water is of low quality. 3c. Possible causes are organic contamination in the water (check the aeration procedure) or dirty glassware (especially the BOD bottles and water storage jug) which has contaminated the water. 3d. The difference in ml readings is not used as a blank correction, but merely as a check on the quality of the 20°C water. 3e. One example of a data sheet is attached.	

Five-Day Biochemical Oxygen Demand (BOD_5)

1. Sample number or other
   identification ______

2. ml of sample used ______

3. % dilution (divide line 2 by 10;
   assumes dilution to 1,000 ml) ______

4. Decimal equivalent of %
   dilution (move the decimal
   point in line 3 two places
   left) ______

5. Initial sample DO in mg/l ______

6. ml of titrant to titrate
   initial blank ______

7. Date of initial DO determination ______

8. Time of initial DO determination ______   (ADDITIONAL COLUMNS
                                                MAY BE ADDED)

9. Sample DO after 5 days in mg/l ______

10. ml of titrant to titrate
    5th day blank ______

11. Date of 5th day DO determination ______

12. Time of 5th day DO determination ______

13. Line 6 minus line 10* ______

14. Line 5 minus line 9 ______

15. BOD_5 in mg/l (line 14 divided
    by line 4) ______

*The number here must not be greater than 0.2.

EFFLUENT MONITORING PROCEDURE: WINKLER DETERMINATION OF DISSOLVED OXYGEN-AZIDE MODIFICATION

1. Analysis Objectives:

The operator will be able to perform a Winkler dissolved oxygen determination, using the azide modification, on a sewage sample.

2. Brief Description of Analysis:

A solution of manganous sulfate is added to the sample. A solution containing sodium hydroxide, sodium iodide and sodium azide is next added. If oxygen is present in the sample, a brown flocculent precipitate forms. If no oxygen is present, a white precipitate forms. Sulfuric acid is then added to the sample, and the precipitate dissolves. The solution is titrated with sodium thiosulfate using starch indicator. At the end point of the titration, the color of the solution changes from pale blue to colorless. The milliliters of sodium thiosulfate used, is equal to the milligrams of dissolved oxygen per liter of sample.

General Description of Equipment Used in the Process

A. Capital Equipment

1. Analytical balance, 200 g. capacity
2. Trip balance, 500 g. capacity
3. Oven, temperature controlable to $\pm\,2^{\circ}C$, large enough to hold a small evaporating dish
4. Refrigerator, large enough to hold three 1 liter bottles
5. Still, or other source of distilled water

B. Reusable

1. Hot plate, large enough to hold a 2 liter Erlenmeyer flask
2. Kemmerer sampler
3. APHA sampler
4. Laboratory apron
5. Safety glasses
6. Brushes (for cleaning glassware)
7. Brush (for cleaning balance)
8. One 1 liter volumetric flask
9. One 300 ml BOD bottle
10. One 1 liter graduated cylinder

11. One 100 ml graduated cylinder
12. One 50 ml graduated cylinder
13. One 10 ml graduated cylinder
14. Six 1 liter glass stoppered bottles
15. One rubber stopper (to fit a 1 liter glass stoppered bottle)
16. One 150 ml glass stoppered bottle
17. One spatula (medium size)
18. One spatula (small size)
19. One 2 liter Erlenmeyer flask
20. One 500 ml wide mouth Erlenmeyer flask
21. One 100 ml pipet
22. One 50 ml pipet
23. One 20 ml pipet
24. One pipet bulb
25. Three 5 ml graduated pipets
26. One desiccator (large enough to hold a small evaporating dish)
27. One evaporating dish (large enough to hold about 10 g. of solid)
28. One 25 ml buret
29. One ring stand
30. One buret clamp
31. One distilled water plastic squeeze bottle
32. One pen or pencil
33. One notebook (for recording data)
34. Eight plastic weighing boats (2–3 inches square)
35. Sponges (for cleaning of laboratory table tops)
36. One stirring rod (about 6 inches long)
37. One powder funnel, about 3 inch diameter

C. Consumable

1. Pottassium dichromate, $K_2Cr_2O_7$
2. Concentrated sulfuric acid, H_2SO_4
3. Soap
   (These three reagents are for cleaning glassware. The quantities needed will
   therefore vary.)
4. 480 g. manganous sulfate tetrahydrate, $MnSO_4.4H_2O$
   400 g. manganous sulfate dihydrate, $MnSO_4.4H_2O$, or 364 g. manganous sulfate monohydrate, $MnSO_4.2H_2O$ may also be used.
5. 500 g. sodium hydroxide, $NaOH$
6. 135 g. sodium iodide, NaI
7. 10 g. sodium azide, NaN_3
8. 10 g. soluble starch
9. 15 ml chloroform, $CHCl_3$

10. 186.15 g. sodium thiosulfate pentahydrate, $Na_2S_2O_3.5H_2O$
11. 6 g. potassium biiodate, $KH(IO_3)_2$
12. 3 g. potassium iodide, KI
13. 10 ml concentrated sulfuric acid, H_2SO_4

The quantities given in 4 through 11 above will suffice for approximately 450 determinations of dissolved oxygen. Depending on usage, smaller quantities may be prepared.

All reagents should be of high quality. Different chemical manufacturers may have different ways of indicating a high quality reagent. While no endorsement of one chemical manufacturer over another is intended, the following are some designations used in four chemical catalogs to indicate high quality reagents.

Catalog	Designations
Thomas	Reagent, ACS, Chemically Pure (CP)
Matheson, Coleman & Bell	Reagent, ACS
Curtin Matheson Scientific, Inc.	Primary Standard, ACS, AR
Fisher	Certified, ACS

EFFLUENT MONITORING PROCEDURE: Winkler Determination of Dissolved Oxygen-Azide Modification

OPERATING PROCEDURES	STEP SEQUENCE	INFORMATION/OPERATING GOALS/SPECIFICATIONS	TRAINING GUIDE NOTES
A. Equipment Preparation 1. Cleaning of glassware	1. Clean all glassware and rinse with distilled water.		V.A.1.1
2. Balance preparation	1. Check all balances for cleanliness and proper operation.		
B. Reagent Preparation 1. Manganous sulfate solution	1. Prepare 1 liter of solution containing 480 g. of manganous sulfate tetratrydrate, $MnSO_4 \cdot 4H_2O$	1a. Unless otherwise specified, solutions should be stored in glass stoppered bottles. 1b. Unless otherwise specified, the term water means distilled water.	
2. Alkaline iodide azide solution	1. Dissolve 500 g. of sodium hydroxide, NaOH, in 500 ml of water.	1a. Caution: heat is generated	
	2. Cool the solution to room temperature.		
	3. Dissolve 135 g. of sodium iodide, NaI, in 200 ml of water.		
	4. Dissolve 10 g. of sodium azide, NaN_3, in 40 ml of water.		
	5. Combine the three solutions and dilute to 1 liter.	5a. This solution should be stored in a glass bottle fitted with a rubber stopper.	

EFFLUENT MONITORING PROCEDURE: Winkler Determination of Dissolved Oxygen-Azide Modification

OPERATING PROCEDURES	STEP SEQUENCE	INFORMATION/OPERATING GOALS/SPECIFICATIONS	TRAINING GUIDE NOTES
3. Starch solution	1. Gently boil 1 liter of water on a hot plate.	1a. Proceed with the next steps while the water is heating and boiling.	
	2. Weigh 10 g. of soluble starch.		
	3. Transfer it to a mortar.		
	4. Add about 3 ml of water.		
	5. Grind with a pestle so as to form a thin paste.		
	6. Pour the paste into the boiling water.		
	7. Allow the solution to stand overnight.		
	8. Decant the starch solution into a bottle.	8a. Decanting means to pour slowly so that any solid material will be left behind.	
	9. Add 5 ml of chloroform, $CHCl_3$.	9a. Store in a refrigerator.	
4. Sodium Thiosulfate stock solution 0.75 N (approximate)	1. Boil 1500 ml of water for 3 minutes		
	2. Cool the water to room temperature.		
	3. Weigh 186.15 g. of sodium thiosulfate pentahydrate, $Na_2S_2O_3 \cdot 5H_2O$.		

EFFLUENT MONITORING PROCEDURE: Winkler Determination of Dissolved Oxygen-Azide Modification

OPERATING PROCEDURES	STEP SEQUENCE	INFORMATION/OPERATING GOALS/SPECIFICATIONS	TRAINING GUIDE NOTES
4. Continued	4. Dissolve in 500 ml of the water.		
	5. Dilute to 1 liter with more of the water.		
	6. Add 5 ml of chloroform, $CHCl_3$.	6a. Store in a refrigerator.	
5. Sodium thiosulfate standard titrant, 0.0375N (approximate)	1. Dilute 50.0 ml of the sodium thiosulfate stock solution to 1 liter.	1a. Do not transfer any of the chloroform from the stock solution.	
	2. Add 5 ml of chloroform, $CHCl_3$.	2a. Store in a refrigerator.	
6. Potassium biiodate standard, 0.0375 N	1. Dry 6 g. of potassium biiodate, $KH(IO_3)_2$ at 103°C for 2 hours.		
	2. Cool in a desiccator.		
	3. Prepare 1 liter of a solution containing 4.837 g. of the potassium biiodate.		
	4. Dilute 250.0 ml of this solution to 1 liter.	4a. The N of this solution is 0.0375.	
7. Sulfuric acid, 10% by volume	1. Pour 10 ml of concentrated sulfuric acid, H_2SO_4, into 90 ml of water	1a. Caution: pour the acid slowly. Mix after each addition of 2 ml of the acid.	
	2. Cool the solution to room temperature.		

EFFLUENT MONITORING PROCEDURE: Winkler Determination of Dissolved Oxygen-Azide Modification

OPERATING PROCEDURES	STEP SEQUENCE	INFORMATION/OPERATING GOALS/SPECIFICATIONS	TRAINING GUIDE NOTES
C. Standardization of Sodium Thiosulfate Standard Titrant			
1. Potassium biiodate standard, 0.0375 N	1. Weigh 1-3 g. of potassium iodide, KI.		
	2. Dissolve in 100-150 ml of water.		
	3. Add 10 ml of 10% by volume sulfuric acid.		
	4. Add 20.0 ml of the 0.0375 N potassium biiodate.	4a. Use a volumetric pipette.	
	5. Place the solution in the dark for 5 minutes.		
	6. Add water so as to bring the volume to 300 ml.		
2. Titration	1. Add the approximately 0.0375 N sodium thiosulfate titrant to the solution from a buret until the color changes from red-brown to pale yellow.		
	2. Add 2 ml of starch solution.	2a. A medium blue-pale blue color will form.	
	3. Continue the titration until the color changes from pale blue to colorless.	3a. Ignore any return of blue color.	
	4. Record the ml of sodium thiosulfate used.		

EFFLUENT MONITORING PROCEDURE: Winkler Determination of Dissolved Oxygen-Azide Modification

OPERATING PROCEDURES	STEP SEQUENCE	INFORMATION/OPERATING GOALS/SPECIFICATIONS	TRAINING GUIDE NOTES
3. Calculations	1. Divide the ml of sodium thiosulfate used into 0.75.	1a. The result is the normality of the sodium thiosulfate titrant. It is desirable, but not necessary, that the normality be exactly 0.0375.	II.C.3.1
D. Determination of Dissolved Oxygen			
1. Sample collection	1. If the sample is to be collected from a depth greater than 5 feet, use a Kemmerer sampler.		
	2. If the sample is to be collected from a depth less than 5 feet use an APHA sampler containing a 300 ml BOD bottle.		
	3. If a Kemmerer is used, transfer the sample to a 300 ml BOD bottle. Allow some of the sample to overflow.	3a. Caution: during the sample transfer, do not allow it to splash.	
	4. Carefully insert the stopper of the BOD bottle.	4a. Do not create any air bubbles in the bottle.	
	5. For surface samples, the sample may be collected directly in a 300 ml BOD bottle.	5a. Fill the bottle in such a way that no turbulence is created.	
2. Addition of reagents	1. Remove the stopper and pipette 2.0 ml of manganous sulfate solution into the sample.	1a. Have the tip of the pipette about 1/2 inch below the surface of the liquid. It is desirable, but not necessary, that the normality be 0.0375.	

EFFLUENT MONITORING PROCEDURE: Winkler Determination of Dissolved Oxygen-Azide Modification

OPERATING PROCEDURES	STEP SEQUENCE	INFORMATION/OPERATING GOALS/SPECIFICATIONS	TRAINING GUIDE NOTES
2. Continued	2. Pipette 2.0 ml of alkaline iodide azide solution into the sample.	2a. Have the tip of the pipette about 1/2 inch below the surface of the liquid 2b. A precipitate forms.	
	3. Carefully insert the stopper of the BOD bottle.	3a. Do not create any air bu in the bottle.	
	4. Rinse off the outside of the BOD bottle.	4a. The alkaline iodide azide solution is damaging to the skin.	
	5. Holding the hand over the stopper, invert the BOD bottle slowly 5 times.		
	6. Allow the precipitate to settle.	6a. If it does not settle, wait 2 minutes and proceed.	
	7. Repeat the shaking and settling steps.		
	8. Pipette 2.0 ml of concentrated sulfuric acid into the sample.	8a. The pipette need not be below the surface of the liquid.	
	9. Carefully insert the stopper of the BOD bottle.	9a. Do not create any air bubbles in bottle during this step.	
	10. Rinse off the outside of the BOD bottle.		
	11. Holding the hand over the stopper, invert the BOD bottle slowly five times.	11a. e precipitate will dissolve. 11b. The color of the solution is red-brown if oxygen is present, but colorless if no oxygen is present. If the solution is yellow, a small amount of oxygen is present.	

EFFLUENT MONITORING PROCEDURE: Winkler Determination of Dissolved Oxygen-Azide Modification

OPERATING PROCEDURES	STEP SEQUENCE	INFORMATION/OPERATING GOALS/SPECIFICATIONS	TRAINING GUIDE NOTES
3. Titration	1. Transfer the entire contents of the 300 ml BOD bottle to a wide mouth 500 ml Erlenmeyer flask.		
	2. Add the sodium thiosulfate titrant from a buret until the red-brown color changes to a pale yellow.	2a. If there was little oxygen in the sample and the yellow color was therefore present even before addition of any sodium thiosulfate, the starch should be added immediately.	
	3. Add 2 ml of starch solution.	3a. A medium blue-pale blue color will form.	
	4. Continue the titration until the color changes from pale blue to colorless.	4a. Ignore any return of blue color.	
	5. Record the ml of sodium thiosulfate titrant used.		
4. Calculations	1. Calculate the mg of DO per liter of sample.	1a. mg DO/liter = ml of sodium thiosulfate titrant X N of sodium thiosulfate titrant X 8 x 1000/ml of sample 1b. Since the sample was in a 300 ml BOD bottle, mg DO/liter = ml of sodium tniosulfate X N of sodium thiosulfate X 8 x 1000/300 1c. or, mg DO/liter = ml of sodium thiosulfate X N of sodium thiosulfate x 26.7	

EFFLUENT MONITORING PROCEDURE: Winkler Determination of Dissolved Oxygen-Azide Modification

OPERATING PROCEDURES	STEP SEQUENCE	INFORMATION/OPERATING GOALS/SPECIFICATIONS	TRAINING GUIDE NOTES
4. Continued		1d. If the N of the sodium thiosulfate was exactly 0.0375, then mg DO/liter = ml of sodium thiosulfate x 0.0375 x 8 x 1000/300 1e. or, mg DO/liter = ml of sodium thiosulfate x 1	

EFFLUENT MONITORING PROCEDURE: DETERMINATION OF DISSOLVED OXYGEN USING A DISSOLVED OXYGEN METER

1. Analysis Objectives:

The learner will use the attached EMP to place the Weston and Stack Model 300 Dissolved Oxygen Meter into operation, including electrode cleaning, membrane installation, calibration, and use of the meter to make a dissolved oxygen measurement.

2. Brief Description of Analysis:

The Winkler determination of dissolved oxygen-azide modification is subject to many interferences. In the case of a BOD_5 determination, the problem is minimized to some extent because of sample dilution. If it is felt, however, that appreciable amounts of interferring materials are present, a dissolved oxygen meter should be used.

General Description of Equipment Used in a Process

A. Capital

1. Weston and Stack Model 300 Dissolved Oxygen (DO) Meter with Model A-30 Probe, accessory kit and manufacturers instruction book
2. Still, or other source of distilled water
3. Trip balance, 100 g. capacity

B. Reusable

1. One 100 ml graduated cylinder
2. One 250 ml Erlenmeyer Flask
3. One 100 ml glass stoppered bottle
4. One 200 ml plastic bottle
5. One teaspoon
6. Small blade screwdriver
7. One 250 ml beaker
8. Five cc syringe or eyedropper with tapered end
9. Small pocket knife
10. Four 300 ml BOD bottles
11. Equipment for performing a Winkler DO determination-azide modification, see

---

Mention of a particular brand name does not constitute endorsement by the U.S. Environmental Protection Agency.

EMP CH.O.EMP.1.8.74, Winkler Determination of Dissolved Oxygen-Azide Modification

C. Consumable

1. Potassium iodide, KI, 50 g.
2. Sodium sulfite, $Na_2S_2O_3$, 25 g.
3. Sodium hydroxide, NaOH, 10 g.
4. One rubber band
5. Paper towels
6. Silicone lubricant
7. Source of distilled water
8. One 1 inch long piece of scotch tape
9. Reagents for performing a Winkler DO determination-azide modification, see EMP CH.O.EMP.1.8.74, Winkler Determination of Dissolved Oxygen-Azide Modification
10. Potassium dichromate, $K_2Cr_2O_7$
11. Concentrated sulfuric acid, H_2SO_4
12. Soap
   (Items 10, 11, and 12 are for cleaning galssware, The quantities will therefore vary.)
13. Brushes (for cleaning glassware)
14. Brush (for cleaning balance)
15. Sponges (for cleaning of laboratory table tops)

All reagents should be of high quality. Different chemical manufactureres may have different ways of indicating a high quality reagent. While no endorsement of one chemical manufacturer over another is intended, the following are some designations used in four chemical catalogs to indicate high quality reagents.

Catalog	Designations
Thomas	Reagent, ACS, Chemically Pure (CP)
Matheson, Coleman & Bell	Reagent, ACS
Curtin Matheson Scientific, Inc.	Primary Standard, ACS, AR
Fisher	Certified, ACS

FRONT VIEW OF METER

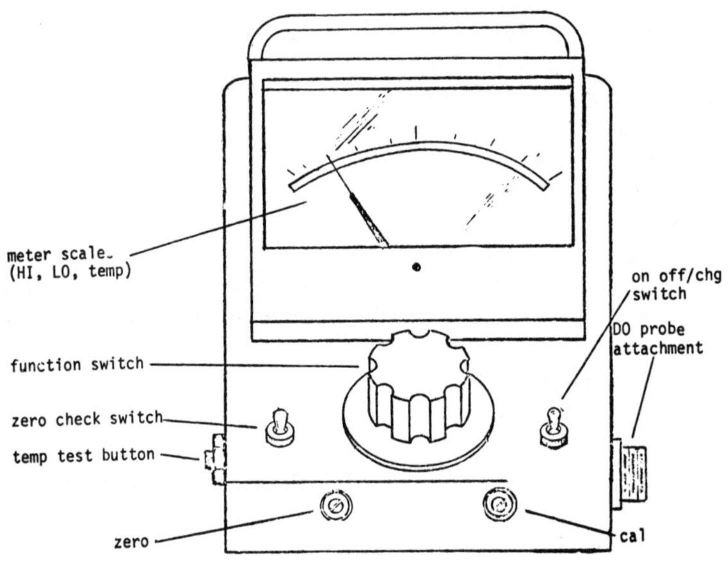

REAR VIEW OF METER

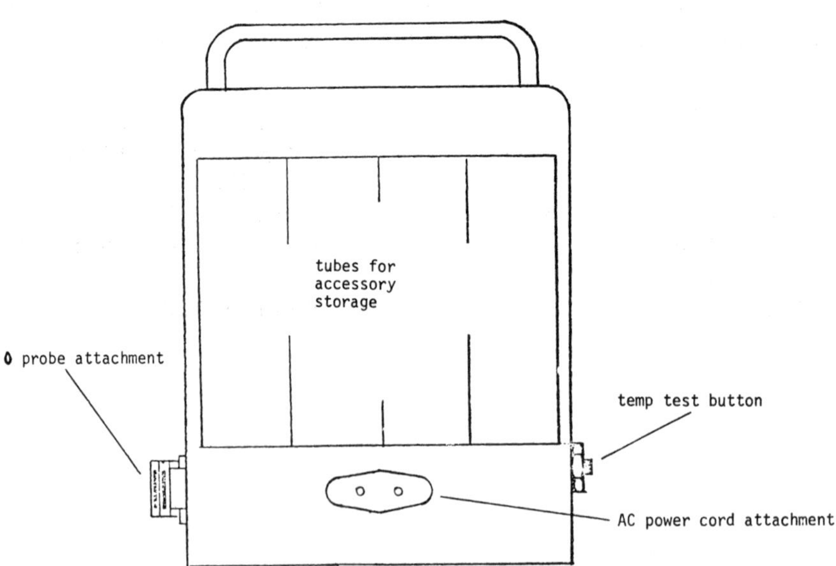

CUT-AWAY VIEW OF PROBE, SHELL, AND STIRRING MECHANISM

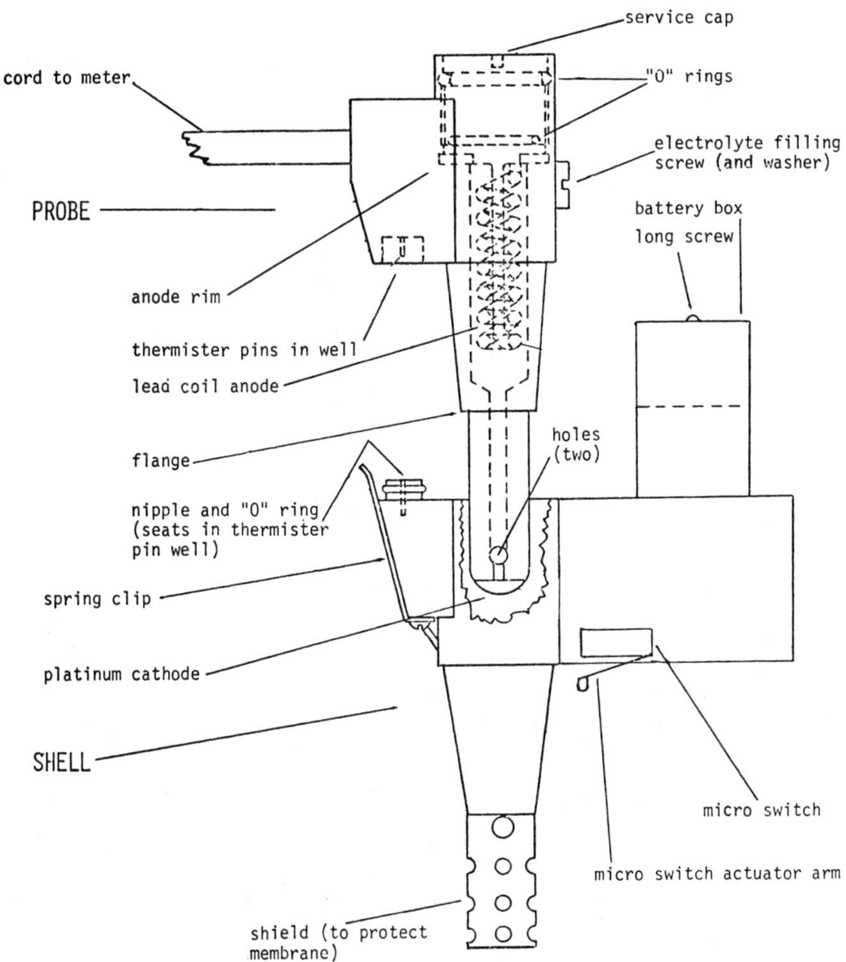

EFFLUENT MONITORING PROCEDURE: Determination of Dissolved Oxygen Using a Dissolved Oxygen Meter

OPERATING PROCEDURES	STEP SEQUENCE	INFORMATION/OPERATING GOALS/SPECIFICATIONS	TRAINING GUIDE NOTES
A. Equipment Preparation 1. Glassware	1. Clean all glassware and rinse with distilled water.		V.A.1.1
2. Balance inspection	1. Check all balances for cleanliness and proper operation.		
B. Reagent Preparation 1. Electrolyte solution	1. Weigh 50 g. of potassium iodide, KI.		
	2. Weigh 0.1 g. of sodium sulfite.		
	3. Dissolve the two solids in 100 ml of water.	3a. Unless otherwise specified, the term water means distilled water. 3b. Unless otherwise specified, solutions should be stored in glass stoppered bottles.	
	4. Store the electrolyte in a small bottle.		
2. Sodium hydroxide solution	1. Weigh 10 g. of sodium hydroxide, NaOH.		
	2. Dissolve it in 90 ml of water.		
	3. Store the solution in a small plastic bottle.		
3. Sodium sulfite solution	1. Measure 1 teaspoon of sodium sulfite, $Na_2S_2O_3$.		
	2. Dissolve it in 500 ml of tap water.	2a. Prepare this solution just prior to use.	

RATING PROCEDURES	STEP SEQUENCE	INFORMATION/OPERATING GOALS/SPECIFICATIONS	TRAINING GUIDE NOTES
C. Equipment Preparation 1. Battery check - Weston and Stack Model 300 Dissolved Oxygen (DO) Meter	1. Check the power cord attached to the rear of the meter.	1a. The meter is portable and the cord should be plugged in to recharge the nickle cadmium battery during storage periods.	
	2. Turn the function switch to the temperature position.		
	3. ....le pressing the temp test button (left side of meter), adjust the temp adj screw (right side of meter) to read 50°C (bottom scale on front of meter).		
	4. Turn the function switch to the transit position.		
2. Battery check-DO probe stirring mechanism	1. Remove the long screw from the top of the battery box.		
	2. Remove the top half of the battery box.		
	3. Insert two size AA 1-1/2 volt batteries (provided with the instrument) into the battery box.	3a. One should be upright, the other upside down.	
	4. Place the top on the battery box.	4a. Note that there is a tip and a hole on the bottom of the top half of the battery box. These fit into a hole and a tip on the top of the bottom half of the battery box.	
	5. Insert the long screw into the top of the battery box.		
	6. Screw it down.		
	7. Using your finger, close the micro switch actuator arm on the side of the probe.	7a. The probe stirring mechanism will start.	

EFFLUENT MONITORING PROCEDURE: Determination of Dissolved Oxygen Using a Dissolved Oxygen Meter

OPERATING PROCEDURES	STEP SEQUENCE	INFORMATION/OPERATING GOALS/SPECIFICATIONS	TRAINING GUIDE NOTES
2. Continued	8. Loosen the long screw.		
	9. Raise the top half of the battery box about half inch.		
	10. Wrap a rubber band around the battery box.	10a. The rubber band should be placed in such a way that the two halves of the battery box are kept apart. 10b. This will keep the stirring mechanism from operating when the probe is not in use. 10c. Do not remove the rubber band and tighten the long screw until the meter is to be calibrated or measurements are to be made.	
3. Cathode check-DO probe	1. Unfasten the spring clip and remove the probe from the shell.		
	2. Examine the platinum cathode at the end of the probe.	2a. It should be free of dirt. 2b. If it is not, wipe it briskly with a paper towel, or coarse piece of cloth.	
4. Anode check-DO probe	1. Remove the service cap on top of the probe.		
	2. Examine the two black "O" rings.	2a. They should be free of dirt.	
	3. Coat the two "O" rings with a very thin layer of silicone lubricant.		
	4. Invert the probe over a table top.	4a. The lead coil anode should drop out. 4b. If it does not, tap the probe lightly on the table top.	

EFFLUENT MONITORING PROCEDURE: Determination of Dissolved Oxygen Using a Dissolved Oxygen Meter

OPERATING PROCEDURES	STEP SEQUENCE	INFORMATION/OPERATING GOALS/SPECIFICATIONS	TRAINING GUIDE NOTES
4. Continued	5. Examine the anode.	5a. It should be free of dirt and corrosion. Yellow colored corrosion is common.	
	6. If corrosion is present, soak the anode in the sodium hydroxide solution.	6a. A few minutes soaking should suffice. 6b. Very minute amounts of corrosion will not cause problems.	
	7. Rinse the anode thoroughly with tap water.		
	8. Rinse the anode thoroughly with distilled water.		
	9. Examine the rim, inside of the probe, on which the anode sits.	9a. It should be free of dirt and corrosion. Yellow colored corrosion is common.	
	10. If corrosion is present, scrape it away using the blade of a small screwdriver.	10a. A swab dipped in the sodium hydroxide solution will also remove the corrosion.	
	11. Rinse the rim and interior of the probe thoroughly with tap water and then with distilled water.		
5. Thermister contact check	1. Invert the probe.		
	2. Examine the two thermister pins in the small well.	2a. They should be free of dirt and corrosion. 2b. If they are corroded, gently scrap them, using the blade of a small screwdriver.	
	3. Examine the "O" ring around the nipple which seats in the thermistor pin well.	3a. It should be free of dirt.	

EFFLUENT MONITORING PROCEDURE: Determination of Dissolved Oxygen Using a Dissolved Oxygen Meter

OPERATING PROCEDURES	STEP SEQUENCE	INFORMATION/OPERATING GOALS/SPECIFICATIONS	TRAINING GUIDE NOTES
5. Continued	4. Coat it with a very thin layer of silicone lubricant.		
6. Membrane installation	1. Select a 1 mil membrane (furnished with the instrument).	1a. A mil is 0.001 inch. 1b. One-half mil membranes are sometimes used. They respond faster, but are more fragile.	
	2. examine it in bright light for holes.	2a. If any are seen, discard the membrane.	
	3. Hold the probe upside down.		
	4. Lay the square membrane over the platinum cathode.	4a. The cathode should be in the center of the square.	
	5. Fold the membrane back over the probe.		
	6. Loop a small rubber band (furnished with the instrument) around the probe three times just above the flange.	6a The rubber band should hold the membrane snugly, but not too tightly. 6b. Two turns of the rubber band may suffice in some cases.	
	7. Gently pull on the loose edges of the membrane.	7a. There should be no folds in the membrane over the platinum cathode. 7b. The pulling should not, however, cause tearing of the membrane.	
	8. Place a short piece of scotch tape over the well containing the two thermistor pins.	8a. This will prevent moisture from getting into the thermistor pin well while the probe is being filled with electrolyte.	
	9. Remove the electrolyte filling screw and washer.		
	10. Hold the probe in a verticle position, with a finger tip held over the electrolyte filling well.	10a. The cathode should point down.	

EFFLUENT MONITORING PROCEDURE: Determination of Dissolved Oxygen Using a Dissolved Oxygen Meter

OPERATING PROCEDURES	STEP SEQUENCE	INFORMATION/OPERATING GOALS/SPECIFICATIONS	TRAINING GUIDE NOTES
6. Continued	11. Pour electrolyte solution into the service cap opening.	11a. Fill the probe interior almost to the top.	
	12. With a twisting motion, slide the rubber band and membrane down about 1/2 inch.	12a. The pocket formed by the membrane will fill with electrolyte.	
	13. With a similar motion, slide the rubber band and membrane back up to its original position.	13a. There should be no wrinkles in the part of the membrane lying across the platinum cathode.	
	14. Loop a second rubber band around the membrane about half-way between the first rubber band and the two holes near the platinum cathode.	14a. Wrap the rubber band as tightly as possible, one turn next to another. 14b. There should still be no wrinkles in the part of the membrane lying across the platinum cathode. 14c. It may be awkward to keep a finger tip over the electrolyte filling well. 14d. No problem is created if some electrolyte is lost, it will be replaced later.	
	15. Carefully drop the lead anode into place.		
	16. Screw in the service cap about half-way.		
	17. Hold the probe in a horizontal direction, electrolyte filling hole up.		
	18. Remove the finger tip from the electrolyte filling well.	18a. Some electrolyte will probably run out.	
	19. Using a syringe or eyedropper with a tapered end, add additional electrolyte through the filling hole.	19a. Gently rock the probe back and forth after each addition so as to dislodge air bubbles. 19b. Tap the sides of the probe with a pen or pencil to ensure bubble escape.	

<u>EFFLUENT MONITORING PROCEDURE</u>: Determination of Dissolved Oxygen Using a Dissolved Oxygen Meter

OPERATING PROCEDURES	STEP SEQUENCE	INFORMATION/OPERATING GOALS/SPECIFICATIONS	TRAINING GUIDE NOTES
6. Continued	20. Tighten the service cap.		
	21. Check for completeness of electrolyte filling.		
	22. Replace the electrolyte filling screw and washer.		
	23. Ensure that no air bubbles have formed under the membrane covering the platinum cathode.	23a. If there are any, the membrane should be removed and reinstalled.	
	24. Cut the membrane around the probe between the two rubber bands.		
	25. Cut away the first rubber band.		
	26. Remove the excess membrane.		
	27. Carefully rinse the outside of the probe and membrane with water.		
	28. Gently shake off the water.		
	29. Remove the piece of tape covering the thermistor pin well.		
	30. Carefully place the probe back in the shell.	30a. Be cautious not to tear the membrane. 30b. Make sure the spring clip is properly closed.	
	31. Place the probe in a 300 ml BOD bottle full of water.		

EFFLUENT MONITORING PROCEDURE: Determination of Dissolved Oxygen Using a Dissolved Oxygen Meter

OPERATING PROCEDURES	STEP SEQUENCE	INFORMATION/OPERATING GOALS/SPECIFICATIONS	TRAINING GUIDE NOTES
6. Continued	32. With your eyes at the same level as the end of the probe, look carefully at the end of the probe for about one minute.	32a. If there are any holes in the membrane the leaking electrolyte will be seen in the water, and the membrane must be replaced.	
7. DO meter zeroing	1. Attach the DO probe to the DO meter.	1a. A pliers may be used to assure a snug fit, but be careful not to damage the knurls on the locking collar.	
	2. Rinse the outside of the shell with water.	2a. Do not get water on the micro switch of the stirring mechanism.	
	3. Place the probe in a 300 ml BOD bottle filled with sulfite solution.		
	4. Allow it to stand for 10 minutes.		
	5. Remove the rubber band from the stirring mechanism battery box.		
	6. Screw down the long screw in the top of the battery box.	6a. The stirring mechanism should start since the flared top of the BOD bottle closes the micro switch actuator arm.	
	7. Wait two minutes.		
	8. Turn the on off/chg toggle switch to the on position.		
	9. Turn the function switch to the HI mg/liter position.	9a. The needle on the meter face will move to the right and then slowly drift to the left.	
	10. When the needle reads 1.5 on the top scale, turn the function switch to the LO mg/liter position.	10a. The needle should continue to drift to the left.	

EFFLUENT MONITORING PROCEDURE: Determination of Dissolved Oxygen Using a Dissolved Oxygen Meter

OPERATING PROCEDURES	STEP SEQUENCE	INFORMATION/OPERATING GOALS/SPECIFICATIONS	TRAINING GUIDE NOTES
7. Continued	11. Wait one additional minute.		
	12. Depress the zero check toggle switch and turn the zero screw (bottom front of the instrument) until the needle reads 0 on the middle scale of the meter.	12a. After releasing the toggle switch the needle may slowly drift toward a "true" zero.	
	13. Turn the on off/chg toggle switch to the off/chg position.	13a. In the off/chg position, the nickle cadmium battery is charging when the power cord is attached.	
	14. Turn the function switch to the transit position.		
	15. Loosen the long screw and raise the top half of the battery box.	14a. For the remainder of this EMP, this procedure will be referred to as turning the stirring mechanism off. Lowering the top half of the battery box and tightening the long screw will be referred to as turning the stirring mechanism on.	
	16. Replace the rubber band which separates the two halves of the battery box.		
	17. Remove the probe from the BOD bottle.		
	18. Rinse off the bottom of the probe thoroughly.	17a. Be cautious not to get the micro switch wet. 17b. All traces of the sulfite solution must be removed.	
	19. Place the probe in a BOD bottle filled with water.	18a. To prevent the membrane from drying out, always keep the probe in water when not in use.	
8. DO meter calibration	1. Fill two 300 ml BOD bottles to overflowing with distilled water.	1a. It is essential that the two samples be identical in oxygen content.	

EFFLUENT MONITORING PROCEDURE: Determination of Dissolved Oxygen Using a Dissolved Oxygen Meter

OPERATING PROCEDURES	STEP SEQUENCE	INFORMATION/OPERATING GOALS/SPECIFICATIONS	TRAINING GUIDE NOTES
8. Continued	2. Determine the DO of one of the bottles.	2a. Use the Winkler Method-Azide Modification.	
	3. Remove the probe from the BOD bottle containing the water.		
	4. Rinse the end of the probe with distilled water.		
	5. Place the probe in the second BOD bottle filled with distilled water.		
	6. Turn the on off/chg toggle switch to the on position.		
	7. Turn the stirring mechanism on.		
	8. Turn the function switch to the HI mg/liter position.		
	9. Wait about two minutes.	9a. Because the slower responding 1 mil membrane is being used, about two minutes should be required for the needle reading to stabilize. 9b. With some meters, experience may indicate that a shorter waiting period will suffice.	
	10. Turn the cal screw (bottom front of the meter) until the needle reads (on the upper scale) the same DO value as was obtained from the Winkler titration.		
	11. Turn the stirring mechanism off.		

EFFLUENT MONITORING PROCEDURE: Determination of Dissolved Oxygen Using a Dissolved Oxygen Meter

OPERATING PROCEDURES	STEP SEQUENCE	INFORMATION/OPERATING GOALS/SPECIFICATIONS	TRAINING GUIDE NOTES
8. Continued	12. Turn the on off/chg toggle switch to the off/chg position. 13. Turn the function switch to the transit position. 14. Place the probe back in the BOD bottle of distilled water. The membrane should always be kept wet when not in use.	13a. There is no firm rule about how often to zero and calibrate the Weston and Stack Model 300 DO meter. 13b. A conservative estimate would be to do the calibration daily, and the zeroing every other day in times of frequent use. 13c. Both steps should be performed if the meter has not been used for several days. 13d. After installation of a new membrane, the calibration changes markedly during the first 24 hours, and frequent calibrations are needed during this period.	

APPENDIX F-4

EFFLUENT MONITORING PROCEDURE: MEASUREMENT OF pH

1. Analysis Objectives

2. Brief Description of Analysis*

1. WWTP operator will set up, calibrate and operate portable type pH meter for the pH measurement of wastewater and WWTP effluent. 2. A portable type, battery operated pH meter, equipped with a glass electrode system is used to measure the pH of wastewater treatment plant samples.

General Description of Equipment Used in the Process

A. Capital Equipment

1. pH Meter IL Model 175 PORTO-matic*
   The IL Model 175 PORTO-matic pH meter is a small, solid state, battery operated, portable instrument for the measurement of the pH of aqueous solutions. Manufacturer's Specifications are as follows:
   pH range: 0-14
   pH Scale: 7.2", 1/2% zero centered
   Readability: 0.01 pH
   Electrical Accuracy: better than 0.035 pH
   Drift per Day: less than 0.01 pH
   Battery Life: 2000 hours

B. Reusable

1. Wash bottle, plastic
2. Beakers, 250 ml, 150 ml, 25 ml

C. Consumable

1. Buffer Solution pH 4
2. Buffer Solution pH 9
3. Buffer Solution pH 6.9
4. Buffer Solution pH 7.4
5. Saturated KC1 Solution

---

*Standard Methods for the Examination of Water and Wastewater, 13th Ed., 1971, APHA, Washington, D.C., p. 500
*Mention of a specific brand name does not constitute endorsement by the U.S. Environmental Protection Agency.

EQUIPMENT - PORTO-MATIC pH METER

OPERATING CONTROLS FRONT PANEL, 175 PORTO-MATIC pH METER

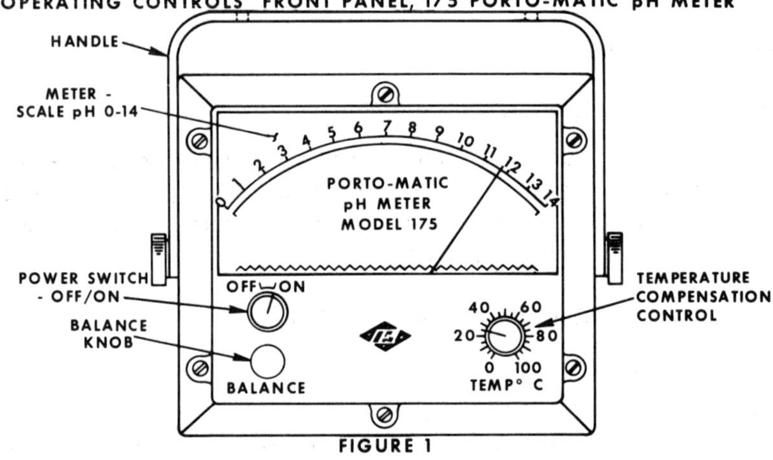

FIGURE 1

REAR PANEL, 175 PORTO-MATIC pH METER

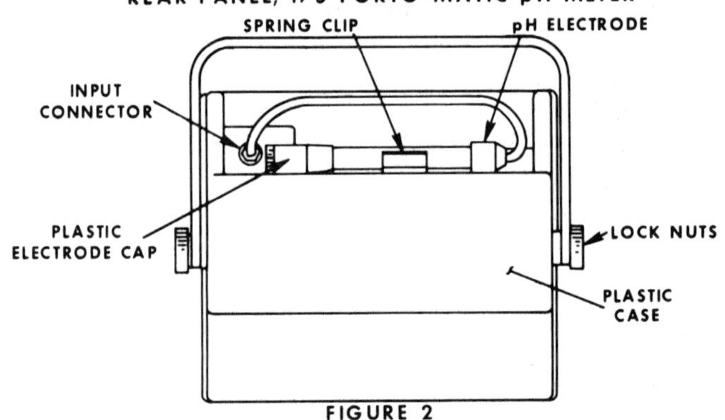

FIGURE 2

EFFLUENT MONITORING PROCEDURE: pH Determination of Wastewater and
Wastewater Treatment Plant Effluents

OPERATING PROCEDURES	STEP SEQUENCE	INFORMATION/OPERATING GOALS/SPECIFICATIONS	TRAINING GUIDE NOTES
A. Instrument Setup	1. Place pH meter on solid surface.		
	2. Remove electrode from rear panel.	See Fig. 2.	I.A (p. 4-14)
	3. Twist the cap at the bottom of the electrode to align the filling holes.	See Fig. 2.	
	4. Check the level and saturation of the KCl solution in the reference chamber of the electrode.		
	5. Place the electrode in a 150 ml beaker containing 100 ml of distilled water.		
	6. Turn ON/OFF switch on.		
B. Meter Calibration	1. Set temperature compensation knob to correspond to temperature of buffer solution to be used for calibration.	1a. Previously prepared standard buffer solution pH 6.9 should be used. Buffer solutions can be prepared from the formulas shown in the attached table.	V.B (p. 4-16)
	2. Transfer 50 ml of pH 6.9 buffer solution into a clean 150 ml beaker.		
	3. Turn meter switch "off"	3a. Meter should be "off" when electrode is out of solution.	

EFFLUENT MONITORING PROCEDURE: pH Determination of Wastewater and
Wastewater Treatment Plant Effluents

OPERATING PROCEDURES	STEP SEQUENCE	INFORMATION/OPERATING GOALS/SPECIFICATIONS	TRAINING GUIDE NOTES
	4. Rinse the electrode with buffer solution and immerse it in the beaker containing the pH 6.9 buffer.	4a. Do not allow bubbles to collect around the ceramic junction of the reference chamber.	V.B (p. 4-16)
	5. Turn meter on.		
	6. Adjust the needle on the meter to read 6.9 by turning the balance knob clockwise or counter-clockwise.	6a. Allow adequate time for the glass electrode to come into equilibrium with the sample (approximately 30 seconds).	
	7. Turn the power switch off.		
	8. Repeat calibration with buffer pH 7.4		
	9. Remove the electrode from the buffer and rinse the electrode three times with distilled water.	9a. Use a squeeze type wash bottle.	
	10. Add distilled water to cap.		
	11. Twist the cap on the bottom of the electrode so that the filling holes are closed and water surrounds the glass membrane.	11a. The pH sensitive membrane dehydrates when removed from water. Dry glass electrodes should be soaked in buffer or water for several hours before use.	
	12. Discard buffer solution.	12a. Never pour used buffer solution back into buffer bottle.	

<u>EFFLUENT MONITORING PROCEDURE</u>: pH Determination of Wastewater and
Wastewater Treatment Plant Effluents

OPERATING PROCEDURES	STEP SEQUENCE	INFORMATION/OPERATING GOALS/SPECIFICATIONS	TRAINING GUIDE NOTES
C. Use of Instrument for pH measurement	1. Adjust temperature compensation knob to the temperature of the unknown solution. 2. Twist open the electrode cap. 3. Immerse the electrode into the unknown. 4. Turn the power switch on. 5. Allow adequate time for the glass electrode to come into equilibrium with the sample (approximately 30 seconds). 6. Determine pH of unknown solution by observation of meter needle on pH scale of instrument. 7. Enter the result on the appropriate report form. Record your value to the nearest 0.1 pH unit. 8. Turn off instrument. 9. Rinse the electrode with distilled water. 10. Add water or buffer to cap prior to closing to prevent dehydration of electrode. 11. Close the cap of the electrode.	5a. Do not allow bubbles to collect around the ceramic junction of the reference chamber. 6a. Swirl probe several times before taking reading. 6b. Take reading with mirror reflection of needle obscured by needle.	

EFFLUENT MONITORING PROCEDURE: pH Determination of Wastewater and
Wastewater Treatment Plant Effluents

OPERATING PROCEDURES	STEP SEQUENCE	INFORMATION/OPERATING GOALS/SPECIFICATIONS	TRAINING GUIDE NOTES
D. Maintenance	1. Check the level and saturation of potassium chloride in the reference chamber of electrode. 2. Keep the glass membrane wet with distilled water when not being used. 3. Do not contaminate standard buffer solution. 4. Turn the instrument off when not in use.	1a. The reference chamber of the pH electrode system should always be kept nearly full with saturated potassium chloride solution. 4a. The pH meter is a battery-operated instrument.	VIII.D (p. 4-16)
E. Trouble Shooting	1. Erratic needle movement: a. Fill the measuring chamber completely b. Soak the external surface of the ceramic plug in warm water. c. Resaturate the reference chamber. 2. No instrumental response when measurement is taken.	1a. Erratic needle movement: Can be caused by bubbles around the ceramic reference junction. Can be caused by contamination or salt crystallization of the reference ceramic plug. Can be caused by unsaturation of the reference chamber. 2a. Exchange electrode or electrodes with new electrodes. 2b. If porous plug of electrode is clogged, take appropriate action to clean. See electrode specification sheet.	

TRAINING GUIDE

Section	Topic
I	Theoretical Concepts
V	Laboratory Analysis
VIII	Maintenance Practices

EFFLUENT MONITORING PROCEDURE: pH Determination of Wastewater and
Wastewater Treatment Plant Effluents

		SECTION I
	TRAINING GUIDE NOTE	REFERENCES/RESOURCES
I	Theoretical Concepts 1. pH General Considerations pH is a term used to describe the intensity of the acid or alkaline condition of a solution. The concept of pH evolved from a series of developments that led to a fuller understanding of acids and alkaline solutions (bases). Acids and bases were originally distinguished by their difference in physical characteristics (acids-sour, bases-soapy feel). In the 18th century it was recognized that acids have a sour taste (vinegar-acetic acid), that they react with limestone with the liberation of a gaseous substance (carbon dioxide) and that neutral substances result from their interaction with alkaline solutions. Acids are also described as compounds that yield hydrogen ions when dissolved in water. And that bases yield hydroxide ions when dissolved in water. The process of neutralization is then considered to be the union of hydrogen (H^+) ions and hydroxyl (OH^-) ions to form neutral water ($H^+ + OH^- \longrightarrow H_2O$). It has been determined that there are 1/10,000,000 grams of hydrogen ions and the same quantity of hydroxyl ions in one liter of pure water. The product of the H^+ and OH^- ions equal a constant value. Therefore, if the concentration of the H^+ ions is increased there is a corresponding decrease in OH^- ions. The acidity or alkalinity, hydrogen ion concentration of a solution is given in terms of pH. The pH scale extends from 0 to 14 with the neutral point at 7.0.	Nebergall, W. H., Schmidt, F. C. and Holtzclaw, Jr., HF., College Chem., 2nd Ed. Heath & Co., Boston, 1963

TRAINING GUIDE NOTE	REFERENCES/RESOURCES
2. Electrode Design About 1925 it was discovered that an electrode could be constructed of glass which would develop a potential related to the hydrogen-ion concentration without interference from most other ions. The glass pH electrode is the nearest approach to a universal pH indicator known at present. It works on the principle of establishing a potential across a pH sensitive, glass membrane whose magnitude is proportional to the difference in pH of the solution separated by this membrane. All glass pH indicating electrodes have a similar basic design. Contained on one side of an appropriate glass membrane is a solution of constant pH. In contact with the other side of this pH sensitive glass is the solution of unknown pH. Between the surfaces of the glass membrane, a potential is established which is proportional to the pH difference of these solutions. As the pH of one solution is constant, this developed potential is a measure of the pH of the other. To measure this potential, a half-cell is introduced into both the constant, internal solution and into the unknown, external solution. These half-cells are in turn connected to your pH meter. The internal reversible half-cell sealed within the chamber of constant pH is almost exclusively a wire of silver-silver chloride. The external reversible half-cell is often silver-silver chloride. If both the internal and external electrodes are combined in a common pH measuring device, the electrode is a combination pH electrode.	Sawyer, C.N., and McCarty, P.L. Chem. for San. Eng. 2nd Ed. McGraw-Hill, NY, 1967 Instruction-Manual IL 175 PORTO-matic pH meter Instrumentation Laboratory, Inc. Lexington, Mass.

	Sections I, V, & VIII
TRAINING GUIDE NOTE	REFERENCES/RESOURCES

	As the function of these half-cells is to provide a steady reference voltage against which voltage changes at the glass pH sensitive membrane can be referred, they must be protected from contamination and dilution by the unknown solutions. This is accomplished by permanently sealing the internal half-cell in a separate chamber which makes electrical contact to the unknown solution through a porous ceramic plug. This ceramic plug allows current to flow, but does not permit exchange of solution to this chamber. Gradually the KCl solution is slowly lost, therefore a filling port is placed in this electrode so that additional saturated potassium chloride can be added.	
V	Laboratory Analysis	
	1. Instrument Calibration	
	The pH balance control, by adding a voltage in series with the pH electrode system, allows the operator to adjust the meter readout to conform to the pH of the calibrating buffer. In general, calibrate the meter in the general range of the unknown solutions. Appropriate buffers can be selected (pH 4.0, 6.8, 7.4 and 10.0). Always set the temperature compensator on the instrument to the temperature of the standard buffer solution.	
	For most accurate analysis the pH of the sample should be determined, and then buffered solutions of a pH above and below the determined pH should be selected to re-calibrate the instrument and the determination of the pH of the sample repeated for a final reading.	
VIII	Maintenance Practices	
	1. The reference chamber of the pH electrode system should always be kept nearly full of saturated KCl solution. Routinely check the level and saturation	

TRAINING GUIDE NOTE	REFERENCES/RESOURCES
of potassium chloride in this reference chamber and add saturated KCl if necessary. 2. The pH sensitive glass membrane dehydrates when removed from water, and thus it is imperative that dry electrodes be soaked in buffer or water for several hours before use. To avoid this break-in period always keep the glass pH sensitive membrane wet between periods of use. 3. The buffers are pH standards; do not contaminate them. 4. The meter is a battery operated instrument. To conserve the battery life, turn the instrument off when not in use.	

Table 144(1): Preparation of pH Standard Solutions

Standard Solution (Molality)	pH at 25 C	Weight of Chemicals Needed per 1,000 ml of Aqueous Solution at 25 C
Primary standards		
Potassium hydrogen tartrate (saturated at 25 C)	3.557	$6.4gKHC_4H_4O_6$*
0.05 potassium dihydrogen citrate	3.776	$11.41gKH_2C_6H_5O_7$
0.05 potassium hydrogen phthalate	4.008	$10.12gKHC_8H_4O_4$
0.025 potassium dihydrogen phosphate + 0.025 disodium hydrogen phosphate	6.865	$3.388gKH_2PO_4$+ + $3.533gNa_2HPO_4$+‡
0.008695 potassium dihydrogen phosphate + 0.03043 disodium hydrogen phosphate	7.413	$1.179gKH_2PO_4$+ + $4.302gNa_2HPO_4$+‡
0.01 sodium borate decahydrate (borax)	9.180	$3.80gNa_2B_4O_7 \cdot 10H_2O$‡
0.025 sodium bicarbonate + 0.025 sodium carbonate	10.012	$2.092gNaHCO_3$ + $2.640gNa_2CO_3$
Secondary standards		
0.05 potassium tetroxalate dihydrate	1.679	$12.61gKH_3C_4O_8 \cdot 2H_2O$
Calcium hydroxide (saturated at 25 C)	12.454	$1.5gCa(OH)_2$*

*Approximate solubility
+Dry chemical at 110-130 C for 2 hr.
‡Prepare with freshly boiled and cooled distilled water (carbon dioxide-free)

APPENDIX F-5

EFFLUENT MONITORING PROCEDURE:

Total Suspended (Non-Filterable) Solids, mg/liter

1. Objective — To determine total suspended (non-filterable solids on a weight (mg/liter) basis.

2. Description of Analysis — A well-mixed sample is filtered through a weighed, standard glass fiber filter disc in a filtration assembly. The filter disc with retained residue is dried in an oven at $103^\circ - 105^\circ$C until a constant weight is obtained. The difference between the weight of the filter disc plus residue (g) and the original weight of the filter disc (g) is divided by the milliliters of sample filtered, then multiplied by 1,000,000. The final result is recorded as total suspended (non-filterable) solids, mg/liter.

Operating Procedures:

A. Prepare the filter disc

60 minutes in oven at 103°–105°C
20–30 minutes in a desiccator

B. Prepare to test the sample.
C. Weigh the filter disc
D. Seat the filter disc
E. Filter the sample
F. Wash down walls of filter apparatus
G. Dry filter disc and residue

    J.1 Clean the filtration equipment

60 minutes in oven at 103°–105°C
20–30 minutes in a desiccator

H. Weigh filter disc and residue
I. Check for complete drying

30 minutes in oven at 103°–105°C
20–30 minutes in a desiccator

---

*Source of Procedure: Methods for Chemical Analysis of Water and Wastes, 1971, Environmental Protection Agency, Analytical Quality Control Laboratory, Cincinnati, Ohio p. 278.

Finish check for complete drying
J.2 Clean filter disc support

K. Calculate total suspended (non-filterable) solids, mg/liter
Report the data

Equipment and Supply Requirements

A. Capital Equipment:

Balance, analytical, capable of weighing to 0.1 mg under a 200 mg load
Oven, drying, for use at $103°-105°C$.
Vacuum source or pump drawing 15 inches mercury

B. Reusable Supplies:

1 Cylinder, graduated, 25 or 50 ml
1 Cylinder, graduated, volume equal to or greater than the volume of sample to
    be filtered. (100 ml is commonly used. For sample volumes less than 10 ml,
    a wide-tip pipet can be used with a pipet bulb to draw sample into pipet.)
1 Desiccator (for storing filter discs on watch glasses, etc.)
1 Flask, suction, with side arm, 1000 ml
1 Hose connection from suction flask to vacuum source
1 Pinchcock clamp to use on hose
1 Filter holder: membrane filter holder assembly or Buchner funnel or Hirsch
    funnel. The filter holder should have a stopper which fits into the mouth of
    the 1000 ml suction flask. Good crucibles may be used — one for each sample
    plus one adapter to hold the crucibles in the mouth of the 1000 ml suction
    flask.
1 Support for filter disc during drying (watch glasses, etc., number depends on
    number of samples). If Gooch crucibles are used, omit this item.
1 Pair Forceps (flat, to handle filter discs)
1 Wash Bottle, squeeze type for distilled water
1 Set Cork Borers

C. Consumable Supplies:

Filter discs, glass fiber, without organic binder, Reeve Angel type 934H or 984H,
    Gelman type A, Whatman GF/C or equivalent. Diameter should be large
    enough to disc will cover openings in the filter holder to be used.
Marking ink to permanently mark glass or porcelain. A marking tool can be used
    instead.
Notebook, bound
Tissues, soft (for balance work)
Water, distilled

EFFLUENT MONITORING PROCEDURE: Total Suspended (Non-Filterable) Solids, mg/liter

OPERATING PROCEDURES	STEP SEQUENCE	INFORMATION/OPERATING GOALS/SPECIFICATIONS	TRAINING GUIDE NOTES
TOTAL SUSPENDED (NON-FILTERABLE) SOLIDS, mg/liter			I (p. 11-25)
A. Preparing the Filter Disc	1. Gather equipment.	1a. See page 6 for list of necessary equipment. The oven should be turned on and set for 103°-105°C temperature. 1b. Filter disc supports (watch glasses, etc.) or Gooch crucibles should have permanent identification marks. 1c. Be sure equipment is clean.	V.A.1b (p. 11-27) V.A.1c (p. 11-27)
	2. Place filter holder with stopper or adapter into the suction flask.	2a. Twist, pressing downward for air-tight fit.	
	3. Attach hose.	3a. From side arm of suction flask to the vacuum source.	
	4. Pick up a filter disc.	4a. Using forceps.	
	5. Place filter disc on the filter holder.	5a. Wrinkled surface of filter disc facing upward. 5b. Disc should cover all openings in filter holder.	
	6. Apply vacuum.	6a. Gradually, to seat the filter disc. A pinchcock clamp on the vacuum hose can be used to regulate application of vacuum. 6b. If a membrane filter holder is used, attach funnel now and tighten the collar.	
	7. Measure out about 20 ml distilled water.	7a. In a 25 or 50 ml graduated cylinder.	

EFFLUENT MONITORING PROCEDURE: Total Suspended (Non-Filterable) Solids, mg/liter

OPERATING PROCEDURES	STEP SEQUENCE	INFORMATION/OPERATING GOALS/SPECIFICATIONS	TRAINING GUIDE NOTES
A. Preparing the Filter Disc (Continued)	8. Pour the 20 ml distilled water on to the disc.	8a. Vacuum still being applied. 8b. To rinse off the filter disc. 8c. If fibers of disc form a lumpy area, discard the disc and begin again at step 4.	
	9. Measure out about 20 ml distilled water.	9a. In the same graduated cylinder.	
	10. Pour this second 20 ml amount of distilled water on to the disc.	10a. Vacuum still being applied. 10b. A second rinse for the disc.	
	11. Measure out about 20 ml distilled water.	11a. In the same graduated cylinder.	
	12. Pour this third 20 ml amount of distilled water on to the disc.	12a. Vacuum still being applied. 12b. A third rinse for the disc.	
	13. Continue vacuum application.	13a. For 2 minutes to remove all traces of water. 13b. If a membrane filter holder is used, loosen the collar and remove the funnel.	
	14. Turn off vacuum.	14a. Break vacuum by pushing upward on the rubber adapter or stopper until air can enter the flask.	
	15. Loosen the filter disc from the filter holder.	15a. If a Gooch crucible is used, omit this step. 15b. If a membrane filter holder is used, use forceps to loosen the disc. Be careful not to damage the disc.	

<u>EFFLUENT MONITORING PROCEDURE</u>: Total Suspended (Non-Filterable) Solids, mg/liter

OPERATING PROCEDURES	STEP SEQUENCE	INFORMATION/OPERATING GOALS/SPECIFICATIONS	TRAINING GUIDE NOTES
A- Preparing the Filter Disc (Continued)	16. Slide the filter disc on to a suitable support.	16a. If a Gooch crucible is used, remove the crucible with the filter disc in it. Wipe the outside with a tissue to remove droplets of water, finger-prints, etc. Do not directly handle the crucible during the procedure. Use tissue, forceps or tongs instead. 16b. If a membrane filter holder is used, use a dry watch glass, etc., to hold the disc. 16c. The filtration assembly can be left as is for future use.	
	17. Put disc (on support) into an oven.	17a. To dry at 103°-105°C. 17b. For 30 minutes in a mechanical convection oven. 17c. For 60 minutes in a gravity convection oven. 17d. Note: Do not open oven door during drying period.	
	18. Remove disc (on support) from oven.	18a. With tongs or gloves, etc.	
	19. Put disc (on support) into desiccator.	19a. Desiccant must be dry. 19b. Desiccator should be air-tight with enough room so disc supports do not touch each other or the side of the desiccator.	V.A.19a. (p. 11-27)
	20. Store disc in desiccator until needed.	20a. Disc and support should be cooled to room temperature before weighing--20 to 30 minutes.	

<u>EFFLUENT MONITORING PROCEDURE</u>: Total Suspended (Non-Filterable) Solids, mg/liter

OPERATING PROCEDURES	STEP SEQUENCE	INFORMATION/OPERATING GOALS/SPECIFICATIONS	TRAINING GUIDE NOTES
B. Preparing to Test the Sample	1. Assemble filtering equipment except for filter disc.	1a. Equipment list is on page 6. 1b. The filtering assembly used to prepare the disc (rinsing it) can be re-used at this time. 1c. Rinse water can remain in suction flask. 1d. Filter holder or Gooch crucible adapter should be tightly in mouth of suction flask. 1e. The oven should be turned on and set for 103°-105°C temperature.	
	2. Record the sample identification information.	2a. Sample should be at hand before continuing with this test. 2b. Use a laboratory notebook. 2c. Record "identification", "type", "date and time collected", and name of "sample collector."	VII.B.2a (p. 11-28) IX.B.2b (p. 11-31) IX.B.2c (p. 11-31)
C. Weighing the Filter Disc	1. Bring forceps, record book and pen to balance table.	1a. Use an analytical balance.	
	2. Zero the balance.		
	3. Remove filter disc (on support) from desiccator.	3a. If a Gooch crucible is being used, use a tissue, forceps or tongs to remove it from the desiccator. It should contain a rinsed, dried filter disc.	
	4. Record filter disc identification.	4a. Gooch crucible number or watch glass number. (Examples: C-12, WG-1) 4b. In laboratory notebook. 4c. In column of the sample for which this disc will be used. 4d. Labeled "filter identification."	V.C.4a (p. 11-27) IX.C.4b (p. 11-31) IX.C.4d (p. 11-31)

<u>EFFLUENT MONITORING PROCEDURE</u>: Total Suspended (Non-Filterable) Solids, mg/liter

OPERATING PROCEDURES	STEP SEQUENCE	INFORMATION/OPERATING GOALS/SPECIFICATIONS	TRAINING GUIDE NOTES
C. Weighing the Filter Disc (Continued)	5. Place filter disc on balance pan.	5a. If a Gooch crucible is being used, use a tissue, forceps or tongs to place it on the balance pan. 5b. If a membrane filter holder is being used, use forceps to slide the filter disc from the storage support (watch glass, etc.) on to the pan.	
	6. Weigh the filter disc.	6a. To four decimal places. 6b. For Gooch crucibles, you can save weighing time by keeping a list of the numbered crucibles with their approximate weights so you have a beginning weight for this operation.	
	7. Record the weight.	7a. In laboratory notebook, 7b. In column of the sample for which this disc will be used. 7c. Labeled "weight of filter (g)." If Gooch crucibles are used, this is the weight of the crucible containing a filter disc.	
	8. Remove the filter disc from the balance pan.	8a. If a Gooch crucible is being used, use tissue, forceps or tongs to remove crucible containing filter disc. 8b. If a membrane filter holder is being used, use forceps to slide the filter disc from the pan on to its storage support (watch glass, etc.).	
	9. Return all weights on the balance to zero position.		

EFFLUENT MONITORING PROCEDURE: Total Suspended (Non-Filterable) Solids, mg/liter

OPERATING PROCEDURES	STEP SEQUENCE	INFORMATION/OPERATING GOALS/SPECIFICATIONS	TRAINING GUIDE NOTES
D. Seating the Filter Disc	1. Slide the filter disc on to the filter holder held in the mouth of the suction flask.	1a. If a Gooch crucible is used, put crucible and disc into the Gooch crucible adapter. 1b. If a membrane filter holder is used, place the wrinkled surface of the disc facing upward on the filter holder. 1c. If a series of funnel type filter assemblies are used, be sure to write the filter disc identification number on the corresponding funnel or flask.	
	2. Apply vacuum.	2a. Gradually, to help seat filter disc. A pinchcock clamp on the vacuum hose can be used to regulate application of vacuum. 2b. If a membrane filter holder is being used, attach funnel now and tighten the collar.	
	3. Pour about 5 ml distilled water on to the filter disc.	3a. Vacuum still being applied. 3b. Can use squeeze bottle of distilled water and estimate volume. 3c. Wetting helps seat filter against holder.	
	4. Leave vacuum on.		
E. Filtering the Sample	1. Record date and time.	1a. In laboratory notebook. 1b. In column of the sample to be filtered. 1c. Labeled "Date and Time Analysis Began."	IX.E.1a (p. 11-31) IX.E.1c (p. 11-31)
	2. Select the volume of sample to be filtered.	2a. 100 ml of sample is a commonly used volume. 2b. CAUTION: Too much residue on the filter will entrap water and may require prolonged drying. If suspended solid concentration in the sample is obviously great, choose a less-than-100 ml volume of well mixed sample.	VII.E.2b (p. 11-29)
	3. Shake the sample.	3a. So portion used is representative of all the sample.	

EFFLUENT MONITORING PROCEDURE: Total Suspended (Non-Filterable) Solids, mg/liter

OPERATING PROCEDURES	STEP SEQUENCE	INFORMATION/OPERATING GOALS/SPECIFICATIONS	TRAINING GUIDE NOTES
E. Filtering the Sample (Continued)	4. Immediately measure out the selected volume.	4a. Using a graduated cylinder (use a wide tip pipet for volumes less than 10 ml). 4b. Measure rapidly since solids may settle in the sample container while you are filling the cylinder. 4c. If you pour the sample to above the graduations, pour that sample back into the bottle and begin again at Step 3.	
	5. Pour the sample on to the filter disc in the filtration assembly.	5a. You should filter all the sample you measure out because you should rinse remaining, settled solids out of the cylinder and on to the filter disc. 5b. If a series of samples are being filtered, be sure you filter each sample through the filter disc you weighed and designated for that sample on the lab data sheet.	
	6. Rinse any sample left in the cylinder on to the filter disc.	6a. With distilled water. 6b. As required. 6c. If suspended solid concentration on the filter disc is obviously small, measure additional volumes of well-mixed sample and filter these, rinsing the cylinder each time.	
	7. Leave suction on.		
	8. Record the total ml of sample filtered.	8a. In laboratory notebook. 8b. In column of the sample filtered. 8c. Labeled "ml Sample Filtered."	

EFFLUENT MONITORING PROCEDURE: Total Suspended (Non-Filterable) Solids, mg/liter

OPERATING PROCEDURES	STEP SEQUENCE	INFORMATION/OPERATING GOALS/SPECIFICATIONS	TRAINING GUIDE NOTES
F. Washing Walls of Filter Apparatus	1. Rinse walls of filter holder with about 10 ml distilled water.	1a. A squeeze bottle of distilled water can be used. Estimate the 10 ml volume. 1b. Otherwise, use a graduate and direct the rinse onto the walls. 1c. Suction should be applied.	
	2. Allow time for complete drainage.		
	3. Rinse walls of filter holder with another 10 ml distilled water.	3a. See information above for F.1.	
	4. Allow time for complete drainage.		
	5. Rinse walls a third time with about 10 ml distilled water.	5a. See information above for F.1.	
	6. Continue vacuum application.	6a. For two minutes to remove all traces of water. 6b. If a membrane filter holder is used, loosen the collar and remove the funnel.	
	7. Turn off vacuum.	7a. Break vacuum by pushing upward on the rubber adapter or stopper until air can enter the flask.	

EFFLUENT MONITORING PROCEDURE: Total Suspended (Non-Filterable) Solids, mg/liter

OPERATING PROCEDURES	STEP SEQUENCE	INFORMATION/OPERATING GOALS/SPECIFICATIONS	TRAINING GUIDE NOTES
G. Drying the Filter Disc and Residue	1. Loosen the filter disc from the filter holder.	1a. If a Gooch crucible is used, omit this step. 1b. If a membrane filter holder is used, use forceps to loosen the filter disc.	
	2. Slide the filter disc plus residue on to its support.	2a. If a Gooch crucible is used, remove the crucible with the filter disc in it. Wipe the outside with a tissue to remove droplets of water, fingerprints, etc. before drying. 2b. If a membrane filter holder is used, slide the filter disc on to the same marked watch glass you used earlier for its support.	
	3. Put disc (on support) into an oven.	3a. To dry at 103°-105°C. 3b. For 30 minutes in a mechanical convection oven. 3c. For 60 minutes in a gravity convection oven. 3d. NOTE: Do not open oven door during drying period. 3e. NOTE: While solids are in drying oven, do "J. Cleaning the Equipment, Step 1."	VII.G.3b (p. 11-29) VII.G.3c (p. 11-29)
	4. Remove disc (on support) from oven.	4a. With tongs or gloves, etc. 4b. Let oven turned on and set for 103°-105°C temperature.	
	5. Put disc (on support) into desiccator.	5a. Desiccant must be dry. 5b. Desiccator should be air-tight with enough room so disc supports do not touch each other or the side of the desiccator.	V.G.5a (p. 11-27)
	6. Allow time for disc to cool to room temperature.	6a. Twenty to 30 minutes.	

EFFLUENT MONITORING PROCEDURE: Total Suspended (Non-Filterable) Solids, mg/liter

OPERATING PROCEDURES	STEP SEQUENCE	INFORMATION/OPERATING GOALS/SPECIFICATIONS	TRAINING GUIDE NOTES
H. Weighing the Filter Disc and Residue	1. Bring forceps, record book and pen to balance table.	1a. Use the same analytical balance you used earlier to weigh the disc.	
	2. Zero the balance.		
	3. Remove filter disc plus residue (on support) from desiccator.	3a. If a Gooch crucible is being used, use a tissue, forceps or tongs to remove it from the desiccator.	
	4. Place filter disc on balance pan.	4a. If a Gooch crucible is being used, use a tissue, forceps or tongs to place it on the balance pan. 4b. If a membrane filter holder is being used, use forceps to slide the filter disc from the storage support (watch glass, etc.) onto the pan.	
	5. Weigh the filter disc plus residue.	5a. To four decimal places. 5b. Use the "weight of the filter" (or of the Gooch crucible with filter) on your Laboratory Data Sheet as a beginning weight.	
	6. Record the weight.	6a. In laboratory notebook, 6b. In column of the sample for which the disc was used. 6c. Labeled "1st weight of filter plus residue (g)." If Gooch crucibles are used, this is the weight of the crucible containing a filter disc with residue.	IX.H.6a (p. 11-31) IX.H.6c (p. 11-31)
	7. Remove the filter disc from the balance pan.	7a. If a Gooch crucible is being used, remove crucible containing filter disc with residue. 7b. If a membrane filter holder is being used, use forceps to slide the filter disc with residue back on to its support (watch glass, etc.)	
	8. Return all weights on the balance to zero position.		

<u>EFFLUENT MONITORING PROCEDURE</u>: Total Suspended (Non-Filterable) Solids, mg/liter

OPERATING PROCEDURES	STEP SEQUENCE	INFORMATION/OPERATING GOALS/SPECIFICATIONS	TRAINING GUIDE NOTES
I. Check for Complete Drying	1. Put disc plus residue (on support) into an oven.	1a. At 103°-105°C. 1b. For 30 minutes. 1c. NOTE: Do not open oven door during drying period.	VII.I. (p. 11-30)
	2. Remove disc (on support) from oven.	2a. Using tongs or gloves, etc. 2b. Let oven turned on and set for 103°-105°C temperature.	
	3. Put disc (on support) into desiccator.	3a. Desiccant must be dry. 3b. Desiccator should be air-tight with enough room so disc supports do not touch each other or the side of the desiccator.	V.I.3a (p. 11-27)
	4. Allow time for disc to cool to room temperature.	4a. Twenty to 30 minutes.	
	5. Bring forceps, record book and pen to balance table.	5a. Use an analytical balance.	
	6. Zero the balance.		
	7. Remove filter disc (on support) from desiccator.	7a. If a Gooch crucible is being used, use a tissue, forceps or tongs to remove it from the desiccator.	
	8. Place filter disc on balance pan.	8a. If a Gooch crucible is being used, use a tissue, forceps or tongs to place it on the balance pan. 8b. If a membrane filter holder is being used, use forceps to slide the filter disc from the storage support (watch glass, etc.) on to the pan.	
	9. Weigh the filter disc plus residue.	9a. To four decimal places. 9b. Use the "1st weight of the filter plus residue (g)" recorded on your data sheet for this sample as a beginning weight.	

EFFLUENT MONITORING PROCEDURE: Total Suspended (Non-Filterable) Solids, mg/liter

OPERATING PROCEDURES	STEP SEQUENCE	INFORMATION/OPERATING GOALS/SPECIFICATIONS	TRAINING GUIDE NOTES
I. Check for Complete Drying (Continued)	10. Record the weight.	10a. In laboratory notebook. 10b. In column of the sample for which the disc was used. 10c. Labeled "2nd weight of filter plus residue (g)." If Gooch crucibles are used, this is the weight of the crucible containing a filter disc with residue.	IX.I.10a (p. 11-31) IX.I.10c (p. 11-31)
	11. Remove the filter disc plus residue from the balance pan.	11a. If a Gooch crucible is being used, remove crucible containing disc with residue. Save this. 11b. If a membrane filter holder is being used, use forceps to slide the filter disc with residue back on to its support (watch glass, etc.). Save this.	
	12. Return all weights on the balance to zero position.		
	13. Find the difference between the 1st and 2nd weights of the filter plus residue.	13a. In laboratory notebook. 13b. In column of the sample for which the disc was used. 13c. Labeled "difference (1st-2nd)."	
	14. Inspect the difference for acceptable agreement of these two weights.	14a. If the weights agree, drying was complete so the procedure is finished. 1) The weights should ideally be constant (the same weight, ± the possible balance error of 0.0001g (0.1 mg). Use the last weight obtained. 2) An acceptable difference between these successive weights is no more than 0.0005g (0.5 mg). In this case, use the last weight obtained for the "final weight of filter plus residue (g)" on line 13 of the Laboratory Data Sheet. (Continued)	II.I.14a.1 (p. 11-26) II.I.14a.2 (p. 11-26)

OPERATING PROCEDURES	STEP SEQUENCE	INFORMATION/OPERATING GOALS/SPECIFICATIONS	TRAINING GUIDE NOTES
I. Check for Complete Drying (Continued)		14b. If the weights do not meet the requirements of agreement, repeat this "Section I: Check for Complete Drying" until you do obtain two successive weights that agree according to a.1) or a.2) above. Use the last weight obtained as the "final weight of filter plus residue (g)" on line 13 of the Laboratory Data Sheet.	
	15. Sign the laboratory data sheet.	15a. In laboratory notebook. 15b. In column for sample(s) you tested. 15c. Labeled "analyst."	
	16. Turn oven off.		
	17. Discard the filter disc plus residue.	17a. Unless there is some reason for saving the solids. 17b. The filter disc support should be cleaned according to "J. Cleaning the Equipment, Step 2."	
J. Cleaning the Equipment	1. Clean the filtration equipment as soon as possible after use (See G.3e).	1a. Membrane filter holder assembly: Leave disc support in suction flask, use squeeze bottle of distilled water, rinse disc support while applying gentle suction. Assembly need not be completely dry before re-use. 1b. Hirsch funnel or Buchner funnel: Leave funnel in suction flask and rinse with distilled water as described above in J.1.1a. 1c. Gooch adapter: Leave in suction flask and rinse the small glass funnel with distilled water (squeeze bottle) while applying gentle suction. Adapter need not be completely dry before re-use. 1d. Suction flask: Remove the rinsed filter holder. Empty the flask through the top (not the side-arm). Rinse it with tap water. Flask need not be completely dry before re-use. (Continued)	

EFFLUENT MONITORING PROCEDURE: Total Suspended (Non-Filterable) Solids, mg/liter

OPERATING PROCEDURES	STEP SEQUENCE	INFORMATION/OPERATING GOALS/SPECIFICATIONS	TRAINING GUIDE NOTES
J. Cleaning the Equipment (Continued)		1e. Graduated cylinders: Rinse with distilled water. These should be dry before re-use. 1f. If stronger cleaning measures are required, use directions given in the Training Guide.	V.J.1f (p. 11-27)
	2. lean the filter disc support .s soon as possible after use. (See I.17.b)	2a. Gooch crucibles: Rinse with distilled water and shake off excess. Crucible need not be completely dry before re-use. 2b. Disc support (watch glass etc.): Rinse with distilled water. Dry completely before re-use. 2c. If stronger cleaning measures are required, use directions given in Training Guide.	V.J.2c (p. 11-27)
K. Calculations	1. Use the following steps to calculate total suspended (non-filterable) solids, mg/liter.	1a. The calculation formula is: Total suspended solids, mg/liter = $$\frac{[(\text{g.wt. filter plus residue}) \text{minus} (\text{g.wt. filter})] \times 1000 \times 1000}{\text{ml sample filtered}}$$ 1b. The "Typical Laboratory Data Sheet" has the steps and an example for doing this calculation. 1c. Numbers used in the examples below are from the example in the third "Sample" column on the "Typical Laboratory Data Sheet."	IX.K.1b (p. 11-31)

<u>EFFLUENT MONITORING PROCEDURE</u>: Total Suspended (Non-Filterable) Solids, mg/liter

OPERATING PROCEDURES	STEP SEQUENCE	INFORMATION/OPERATING GOALS/SPECIFICATIONS	TRAINING GUIDE NOTES
K. Calculations (Continued)			
	2. Subtract the "weight of filter (g)" on line 14 from the "final weight of filter plus residue (g)" on line 13.	2a. Example on data sheet: line 13 - 0.1413 g. line 14 - 0.1293 g. Difference = 0.0120 g. 2b. NOTE: This is the gram weight of the residue which was on the filter disc. 2c. IMPORTANT: This gram weight of the residue (the difference) should be greater than 0.0025 g. If the weight of the residue (the difference) is less than 0.0025 g, you should repeat the procedure and filter a larger volume of the sample so more residue is obtained.	VII.K.2c. (p. 11-30)
	3. Write the difference on line 15 of the data sheet.	3a. This has been done for the example in the third "Sample" column.	
	4. Divide this difference on line 15 by the "ml sample filtered" on line 7 to get a 7 decimal place answer.	4a. Example on data sheet: $\dfrac{\text{line 15}}{\text{line 7}} = \dfrac{0.0120g}{67.0\ ml} = 0.0001791$ g/ml 4b. NOTE: This is the gram weight of residue in each ml of the sample.	
	5. Write this answer on line 16.	5a. This has been done for the example in the third "Sample" column	

EFFLUENT MONITORING PROCEDURE: Total Suspended (Non-Filterable) Solids, mg/liter

OPERATING PROCEDURES	STEP SEQUENCE	INFORMATION/OPERATING GOALS/SPECIFICATIONS	TRAINING GUIDE NOTES
K. Calculations (Continued)	6. Multiply the 7 decimal place answer on line 16 by 1,000,000 (Move the decimal point 6 places to the right).	6a. Example on data sheet: line 16 is 0.0001791 g/ml x 1,000,000 = 179.1 mg/liter 6b. NOTE: This multiplication converts the gram weight of residue per ml to the unit of mg/liter.	
	7. Write this answer on line 17.	7a. This has been done for the example in the third "Sample" column.	
	8. Round off answer on line 17 to to the nearest whole mg.	8a. Example on data sheet: line 17, 179.1 mg/liter becomes: 179 mg/liter	II.K.8a (p. 11-26)
	9. Write this answer on line 18.	9a. This has been done for the example in the third "Sample" column. 9b. Records should be kept in a laboratory notebook.	IX.K.9b (p. 11-31)
L. Reporting Data	1. Report total suspended (non-filterable) solids, mg/liter	1a. On any required record or report sheets.	IX.L.1a (p. 11-31)

EFFLUENT MONITORING PROCEDURE:

Reporting of Self-Monitoring Data

1. Objective: To enable the student to complete the NPDES Discharge Monitoring Report, EPA Form T-40(4-74), or EPA Form 3320-1(10-72).

2. Description of Procedure: Self-monitoring data obtained by a permit holder under the terms of his permit must be reported to the regulatory agency periodically, using the proper NPDES reporting form. The manner in which such data should be reported on EPA Form T-40 is illustrated in this procedure. Additional information required to complete the form is also indicated.

Assumed conditions used to illustrate completion of the form are:

1. Reporting of data on a monthly basis is required.
2. Self-monitoring data developed over a period of one month is as shown in Table 1.
3. Effluent limitations specified in the permit are as shown on Table 2.
4. Monitoring requirements specified in the permit are as shown in Table 2.
5. All required data has been obtained in accordance with permit requirements.

TABLE I
SELF - MONITORING DATA
September 1974

Date	SEWAGE FLOW Treated gpd	RAW INFLUENT BOD_5 mg/1	RAW INFLUENT T.S.S. mg/1	FINAL EFFLUENT BOD_5 mg/1	FINAL EFFLUENT T.S.S. mg/1	FINAL EFFLUENT Fecal Coliform N/100 ml	pH
1	720,100						7.4
2	609,000						7.5
3	326,900	170	171	16	12	350	7.6
4	367,500						7.4
5	323,900						7.5
6	458,500	160	168	15	16	540	7.7
7	571,000						5.4
8	508,600						7.6

continued.

9	146,000	200	200	20	25	180	7.9
10	253,000						7.2
11	406,800						7.1
12	519,200	190	198	20	25	170	7.6
13	328,600						7.5
14	413,100						7.6
15	699,000						8.0
16	708,900	150	180	35	60	220	8.0
17	806,700						9.2
18	714,800						8.0
19	169,100						9.1
20	272,900	170	170	19	19	240	7.5
21	713,200						7.8
22	671,900						7.0
23	761,800	150	186	20	23	110	7.4
24	642,900						7.5
25	314,900						7.4
26	291,600	190	195	20	20	130	7.5
27	240,700						7.4
28	478,900						7.4
29	525,600	190	195	25	25	280	7.6
30	670,100						7.8
Total	14,635,200						
Average	487,800						

TABLE II

EFFLUENT LIMITATIONS AND MONITORING REQUIREMENTS

| | DISCHARGE LIMITATIONS | | | | MINIMUM MONITORING REQUIREMENTS | |
| | Concentration in mg/l | | kg/day (lbs/day) | | | |
EFFLUENT CHARACTERISTICS	Monthly Average	Weekly Average	Monthly Average	Weekly Average	Measurement Frequency	Sample Type
Biochemical Oxygen Demand (5-day)	*30	45	70 (150)		twice weekly	24 hr. composite
Suspended Solids	*30	45	80 (180)		twice weekly	24 hr. composite
pH - standard units	6.0-9.0	(not to be averaged)			twice weekly	grab
Fecal Coliform - organisms/100 ml	200	400	---	---	twice weekly	grab
Flow - mgd	---	---	---	---	daily	recording

* The arithmetic mean of the values for effluent samples measuring biochemical oxygen demand (5-day) and suspended solids collected in a period of 30 consecutive days shall not exceed 15 percent of the arithmetic mean of the values for influent samples collected at approximately the same times during the same period (85 percent removal--minimum).

* Whichever is the more stringent.

EFFLUENT MONITORING PROCEDURE: REPORTING OF SELF-MONITORING DATA

OPERATING PROCEDURES	STEP SEQUENCE	INFORMATION/OPERATING GOALS/SPECIFICATIONS	TRAINING GUIDE NOTES
A. Description of EPA Forms			I.A (P. 27)
B. Identification of Permit Holder and Discharge	1. Enter Name and Address of Permit Holder	1a. In space provided at left of Instructions 1b. May already be entered by permit-issuing authority.	
	2. Enter State in block labelled "ST"	2a. Use standard two-letter postal code 2b. See notes 1a and 1b above.	
	3. Enter Permit number in block for same.	3a. See notes 1a and 1b above.	
	4. Enter discharge number in "Dis" block.	4a. As identified in permit (001, 002, etc.) 4b. See notes 1a and 1b above.	
	5. Enter Discharge code	5a. For municipal wastewater discharges, the number is 4952. 5b. See notes 1a and 1b above.	
	6. Enter Latitude and Longitude of discharge.	6a. If known. 6b. See notes 1a and 1b above.	
	7. Enter reporting period in appropraite blocks.	7a. In this procedure the 30-day period for the month of September is used. 7b. Will be specified in permit. 7c. See notes 1a and 1b above.	

This portion of the report form, completed in accordance with assumed permit conditions, and the data of Table I, is shown in Fig. 1:

EFFLUENT MONITORING PROCEDURE: REPORTING OF SELF-MONITORING DATA

OPERATING PROCEDURES	STEP SEQUENCE	INFORMATION/OPERATING GOALS/SPECIFICATIONS	TRAINING GUIDE NOTES

NATIONAL POLLUTANT DISCHARGE ELIMINATION SYSTEM
DISCHARGE MONITORING REPORT

Form Approved
OMB NO. 158-R0073

City of Noname, Dept. of Environmental Services
184 Any Street
Noname, Anystate, 12345

AN ST · NO1234567 PERMIT NUMBER · 001 DIS · 4952 SIC · 47°20'4" LATITUDE · 98°26'11" LONGITUDE

REPORTING PERIOD: FROM 7 4 0 9 0 1 (YEAR MO DAY) TO 7 4 0 9 3 0 (YEAR MO DAY)

INSTRUCTIONS

1. Provide dates for period covered by this report in spaces marked "REPORTING PERIOD".
2. Enter reported minimum, average and maximum values under "QUANTITY" and "CONCENTRATION" in the units specified for each parameter as appropriate. Do not enter values in boxes containing asterisks. "AVERAGE" is average computed over actual time discharge is operating. "MAXIMUM" and "MINIMUM" are extreme values observed during the reporting period.
3. Specify the number of analyzed samples that exceed the maximum (and/or minimum as appropriate) permit conditions in the columns labeled "No. Ex." If none, enter "0".
4. Specify frequency of analysis for each parameter as No. analyses/No. days. (e.g., "3/7" is equivalent to 3 analyses performed every 7 days.) If continuous enter "CONT."
5. Specify sample type ("grab" or "___ hr. composite") as applicable. If frequency was continuous, enter "NA".
6. Appropriate signature is required on bottom of this form.
7. Remove carbon and retain copy for your records.
8. Fold along dotted lines, staple and mail Original to office specified in permit.

Fig. 1

C. Flow Data.

1. "Quantity" section.

1. Enter minimum and maximum flows during the reporting period in the corresponding spaces on the "reported" line.

1a. Minimum flow of 0.15 MGD on Sept. 9. (Table I)
1b. Maximum flow of 0.81 MGD on Sept. 17. (Table I)

IX.C.1.1 (p. 35)

2. Enter Average flow during reporting period in corresponding space on "reported" line.

2a. Add the flows reported during the reporting period. Divide this total by the number of flows reported.
2b. From Table I, Average = $\dfrac{14{,}635{,}200}{30}$ = 0.49 MGD

3. Enter on "Permit Condition" line the average and maximum daily flows specified in the permit.

3a. If unspecified in Permit, place a dash in the appropriate space or spaces.
3b. May already have been entered by permit-issuing authority.

IX.C.1.3.3a (p. 35)

EFFLUENT MONITORING PROCEDURE: REPORTING OF SELF-MONITORING DATA

OPERATING PROCEDURES	STEP SEQUENCE	INFORMATION/OPERATING GOALS/SPECIFICATIONS	TRAINING GUIDE NOTES
C. Flow Data (cont.)	4. On the "Reported" line, in the "No. Ex" space, enter the number of times during the reporting period that the maximum daily flow specified in the permit was exceeded	4a. If none, enter "0". 4b. If a maximum daily flow is not specified in the permit, place a dash in this space.	
2. "Concentration" section	1. Place dashes in the "Units"space and in the "No. Ex" spaces on the 'Reported" line.	1a. May already be entered by the Permit-issuing authority.	
3. "Frequency of Analysis" Column	1. On the "Reported" line enter the frequency with which flows were measured during the reporting period.	1a. Daily, Weekly, Continuous (Cont.), etc.	
	2. On the "Permit Condition" line enter the frequency of flow measurement as specified in the permit.	2a. May already be entered by Permit-issuing authority.	
4. "Sample Type" Column	1. Enter Dashes on both lines		

This portion of the report form, completed in accordance with assumed permit conditions, and the data of Table I, is shown in Fig. 2:

EFFLUENT MONITORING PROCEDURE: REPORTING OF SELF-MONITORING DATA

OPERATING PROCEDURES	STEP SEQUENCE	INFORMATION/OPERATING GOALS/SPECIFICATIONS	TRAINING GUIDE NOTES

| PARAMETER | | QUANTITY (3 card only) | | | UNITS | NO. EX | CONCENTRATION (4 card only) | | | UNITS | NO. EX | FREQUENCY OF ANALYSIS | SAMPLE TYPE |
		MINIMUM	AVERAGE	MAXIMUM			MINIMUM	AVERAGE	MAXIMUM				
FLOW	REPORTED	0.15	0.49	0.81	MGD	–	******	******	******		–	Cont.	–
	PERMIT CONDITION	******	–	–			******	******	******			Cont.	–

Fig. 2

D. pH Data

1. "Quantity" section

	Step Sequence	Information/Operating Goals/Specifications
	1. On the "Reported" line enter the minimum and maximum pH values occurring during the reporting period.	1a. Although the permit requires that pH be determined twice weekly, a pH was run each day during the month. The report form must be prepared on the basis of all 30 results. 1b. Minimum pH 5.4 on Sept. 7 (Table I) 1c. Maximum pH 9.2 on Sept. 17. (Table I)
	2. On the "Permit Condition" line enter the minimum and maximum pH values specified in the permit.	2a. Permit requires pH to be between 6.0 and 9.0 at all times (Table II) 2b. May already be entered on the form by the permit-issuing authority.
	3. On the "Reported" line, in the "No. Ex" space, enter the total number of times that the pH exceeded the maximum allowed by the permit, and was less than the minimum allowed by the permit.	3a. The maximum permit requirement of 9.0 was exceeded twice - once on Sept. 17, and again on Sept. 19. (Table I) 3b. The pH was less than the minimum permit requirement of 6.0 on Sept. 7. (Table I)

EFFLUENT MONITORING PROCEDURE: REPORTING OF SELF-MONITORING DATA

OPERATING PROCEDURES	STEP SEQUENCE	INFORMATION/OPERATING GOALS/SPECIFICATIONS	TRAINING GUIDE NOTES
D. pH Data (cont.)		3c. Permit requirements were violated a total of 3 times during the reporting period. A "3" should be entered on the Form.	
		3d. If the pH had not exceeded the permit limits, a "0" would be entered.	
2. "Concentration" section	1. Place dashes in the "Units" space and in the "No. Ex" space on the "Reported" line.	1a. May already be entered by Permit-issuing authority.	
3. "Frequency of Analysis" column.	1. Enter frequency of analysis during the reporting period on the "Reported" line.	1a. The actual frequency of analysis is reported. 1b. pH was run each day. The frequency of analysis is reported as 7/7, which indicates that 7 analyses were performed every seven days.	
	2. Enter required frequency of analysis as specified in permit on the "Permit Condition" line.	2a. The permit requires that pH be run twice weekly. (Table II) A 2/7 is entered on this line, which indicates that 2 analyses were to be performed every 7 days. 2b. May already be entered by the Permit-issuing authority.	
4. "Sample Type" column.	1. On the "Reported" line indicate the type of sample on which the analysis was performed.	1a. A grab sample is assumed here. "Grab" is entered.	
	2. On the "Permit Condition" line enter the type of sample specified by the permit.	2a. A grab sample is specified. (Table II) Enter "Grab". 2b. May already be entered by the Permit-issuing authority.	

OPERATING PROCEDURES	STEP SEQUENCE	INFORMATION/OPERATING GOALS/SPECIFICATIONS	TRAINING GUIDE NOTES

D.   pH Data (cont.)

This portion of the report form, completed in accordance with assumed permit conditions, and the data of Table I, is shown in Fig. 3 below:

PARAMETER		QUANTITY (3 card only)				NO EX	CONCENTRATION (4 card only)			UNITS	NO EX	FREQUENCY OF ANALYSIS	SAMPLE TYPE
		MINIMUM	AVERAGE	MAXIMUM	UNITS		MINIMUM	AVERAGE	MAXIMUM				
PH	REPORTED	5.4	******	9.2	STANDARD UNITS	3	******	******	******		-	7/7	Grab
	PERMIT CONDITION	6.0	******	9.0			******	******	******	-		2/7	Grab

Fig. 3

E. BOD_5 Data, Final Effluent.			
1. Computation of Quantities	1. For each reported analytical result, calculate the quantity of BOD_5 discharged in Kg/day.		II.E.1 (p. 30)
2. "Quantity" section	1. On the "Reported" line enter the minimum, average, and maximum quantities discharged over the reporting period, in the appropriate spaces.	1a. Minimum - 11.0 Kg/day 1b. Average - 37.5 Kg/day 1c. Maximum - 94 Kg/day	

OPERATING PROCEDURES	STEP SEQUENCE	INFORMATION/OPERATING GOALS/SPECIFICATIONS	TRAINING GUIDE NOTES
E. BOD$_5$ Data, Final Effluent.(cont.)	2. On the "Permit Condition" line enter the average and maximum quantities permitted to be discharged.	2a. Average BOD$_5$ discharge over a 30-consecutive-day period is specified as 70 Kg/day. (Table II) 2b. A Maximum daily discharge limitation is not indicated in the permit. (Table II) A dash is therefore placed in the "maximum" space. 2c. May already be entered by the Permit-issuing authority.	
	3. In the "No. Ex" column, on the "Reported" line, enter the number of times that the maximum daily discharge of BOD$_5$ (Kg/day) allowed by the permit has been exceeded during the reporting period.	3a. If none, enter a "0". 3b. Since no maximum daily discharge limitation is specified in the permit (Table II), a dash is placed in this space.	
3. "Concentration" section.	1. On the "Reported" line enter the minimum, average, and maximum BOD$_5$ concentrations observed during the reporting period.	1a. Minimum - 15 mg/l 1b. Average - 21 mq/l 1c. Maximum - 35 mg/l	II.E.1 (p. 30)
	2. On the "Permit Condition" line enter the average and maximum concentrations specified in the permit.	2a. Average BOD$_5$ concentration over a 30-consecutive-day period is specified as 30 mg/l. (Table II) 2b. A maximum concentration is not indicated in the permit conditions (Table II). Therefore place a dash in this space. 2c. May already be entered by the Permit-issuing authority.	

EFFLUENT MONITORING PROCEDURE: REPORTING OF SELF-MONITORING DATA

OPERATING PROCEDURES	STEP SEQUENCE	INFORMATION/OPERATING GOALS/SPECIFICATIONS	TRAINING GUIDE NOTES
E. BOD$_5$ Data, Final Effluent.(cont.)	3. In the "No. Ex" column, the "Reported" line, enter the number of times that the maximum concentration allowed by the permit has been exceeded during the reporting period.	3a. If none, enter a "0". 3b. Since no maximum concentration is specified in the permit (Table II) enter a dash in this space.	
4. "Frequency of Analysis" column	1. On the "Reported" line enter frequency of analysis during the reporting period. 2. On the "Permit Condition" line enter required frequency of analysis as specified in permit.	1a. Enter 2/7, indicating that the analysis was performed twice weekly (Table I). 2a. Permit requires analysis twice weekly (Table II). Enter 2/7. 2b. May already be entered by the Permit-issuing authority.	
5. "Sample Type" column	1. On the "Reported" line enter the type of sample on which the analysis was performed. 2. On the "Permit Condition" line enter the type of sample specified in the permit.	1a. Enter "24-Hr. comp." since it is assumed in this procedure that the data has been obtained in accordance with permit requirements. 2a. Enter "24-Hr. comp." (Table II). 2b. May already be entered by the Permit-issuing authority.	

This portion of the Report Form, completed in accordance with assumed permit conditions, and the data of Table I, is shown in Fig. 4:

EFFLUENT MONITORING PROCEDURE: REPORTING OF SELF-MONITORING DATA

OPERATING PROCEDURES	STEP SEQUENCE	INFORMATION/OPERATING GOALS/SPECIFICATIONS	TRAINING GUIDE NOTES

PARAMETER		QUANTITY (3 card only)					CONCENTRATION (4 card only)					FREQUENCY OF ANALYSIS	SAMPLE TYPE
		MINIMUM	AVERAGE	MAXIMUM	UNITS	NO. EX	MINIMUM	AVERAGE	MAXIMUM	UNITS	NO. EX		
BOD_5	REPORTED	11.0	37.5	94	KG/DAY	-	15	21	35	MG/L	-	2/7	24-Hr.Comp
	PERMIT CONDITION	*****	70	-			*****	30	-			2/7	24-Hr.Comp

Fig. 4

F. Percent Removal BOD_5			
1. Computation	1. Calculate the percent BOD_5 removal for each pair of influent and effluent analyses made during the reporting period.		II.F.1 (p. 31)

EFFLUENT MONITORING PROCEDURE: REPORTING OF SELF-MONITORING DATA

OPERATING PROCEDURES	STEP SEQUENCE	INFORMATION/OPERATING GOALS/SPECIFICATIONS	TRAINING GUIDE NOTES
F. Percent Removal BOD$_5$ (cont.)			
2. "Quantity" section	1. On the "Reported" line enter the minimum, average, and maximum percent removals in the appropriate spaces.	1a. Minimum - 77% 1b. Average - 87.9% 1c. Maximum - 91%	
	2. On the "Permit Condition" line enter the minimum and average removals required by the permit.	2a. There is no minimum removal requirement in the assumed permit conditions (Table II). Enter a dash in this space. 2b. Average removal required over a 30-consecutive-day period is 85%. (Table II). 2c. May already be entered by the Permit-issuing authority.	
	3. In the "No. Ex" column, on the "Reported" line, enter the number of times that the minimum percent removal required in the permit was not obtained.	3a. There is no minimum removal requirement in the assumed permit conditions (Table II). Enter a dash in this space.	
3. "Concentration" section	1. Enter dashes in the "Units" space and in the "No. Ex" space on the "Reported" line.	1a. May already be entered by the Permit-issuing authority.	

This portion of the Report Form, completed in accordance with assumed permit conditions, and the data of Table I, is shown in Fig. 5:

EFFLUENT MONITOPING PROCEDURE: REPORTING OF SELF-MONITORING DATA

OPERATING PROCEDURES	STEP SEQUENCE	INFORMATION/OPERATING GOALS/SPECIFICATIONS	TRAINING GUIDE NOTES

PARAMETER		(3 card only) QUANTITY			UNITS	NO. EX	(4 card only) CONCENTRATION			UNITS	NO. EX	FREQUENCY OF ANALYSIS	SAMPLE TYPE
		MINIMUM	AVERAGE	MAXIMUM			MINIMUM	AVERAGE	MAXIMUM				
PERCENT REMOVAL BOD 5	REPORTED	77	87.9	91	%	–	******	******	******	–	–	******	******
	PERMIT CONDITION	–	85	******			******	******	******			******	******

Fig. 5

G. Suspended Solids, Final Effluent			
1. Computation of Quantities	1. For each reported analytical result, calculate the quantity of suspended solids discharged in Kg/day.		II.G.1 (p. 32)
2. "Quantity" section	1. On the "Reported" line enter the minimum, average, and maximum quantities discharged over the reporting period, in the corresponding spaces.	1a. Minimum - 13.8 kg/day 1b. Average - 47.1 kg/day 1c. Maximum - 161 kg/day	

EFFLUENT MONITORING PROCEDURE: REPORTING OF SELF-MONITORING DATA

OPERATING PROCEDURES	STEP SEQUENCE	INFORMATION/OPERATING GOALS/SPECIFICATIONS	TRAINING GUIDE NOTES
G. Suspended Solids, Final Effluent (cont.)	2. On the "Permit Condition" line enter the average and maximum quantities permitted to be discharged.	2a. Average suspended solids discharge over a 30-consecutive-day period is specified as 80 kg/day (Table II) 2b. A maximum daily discharge limitation is not specified in the permit (Table II). A dash is therefore placed in the "maximum" space. 2c. May already be entered by the Permit-issuing authority.	
	3. In the "No. Ex" column, on the "Reported" line, enter the number of times that the maximum daily discharge of suspended solids (kg/day) has been exceeded during the reporting period.	3a. If none, enter a "0". 3b. Since no maximum daily discharge limitation is indicated in the permit (Table II), a dash is placed in this space.	
3. "Concentration" section.	1. On the "Reported" line enter the minimum, average, and maximum suspended solids concentration observed during the reporting period, in the corresponding spaces.	1a. Minimum - 12 mg/l 1b. Average - 25.0 mg/l 1c. Maximum - 60 mg/l	
	2. On the "Permit Condition" line enter the average and maximum concentrations specified in the permit, in the appropriate spaces.	2a. Average suspended solids concentration over a 30-consecutive-day period is specified as 30 mg/l (Table II).	

EFFLUENT MONITORING PROCEDURE: REPORTING OF SELF-MONITORING DATA

OPERATING PROCEDURES	STEP SEQUENCE	INFORMATION/OPERATING GOALS/SPECIFICATIONS	TRAINING GUIDE NOTES
G. Suspended Solids, Final Effluent. (cont.)		2b. Since no maximum concentration is specified in the permit (Table II), enter a dash in this space. 2c. May already be entered by the Permit-issuing authority.	
	3. In the "No. Ex" column, cn the "Reported" line, enter the number of times during the reporting period that the maximum concentration allowed by the permit has been exceeded.	3a. If none, enter a "0". 3b. Since no maximum concentration is specified in the permit (Table II), enter a dash in this space.	
4. "Frequency of Analysis" column	1. On the "Reported" line enter frequency of analysis during the reporting period.	1a. Enter 2/7, indicating that the analysis was performed twice weekly (Table I).	
	2. On the "Permit Condition" line enter required frequency of analysis, as specified in permit.	2a. Permit requires analysis twice weekly (Table II). Enter 2/7. 2b. May already be entered by Permit-issuing authority.	
5. "Sample Type" column.	1. On the "Reported" line enter the type of sample on which the analysis was performed.	1a. Enter "24-Hr. comp." since it is assumed in this procedure that the data has been obtained in accordance with permit requirements.	
	2. On the "Permit Condition" line enter the type of sample specified in the permit.	2a. Enter "24-Hr. comp." (Table II). 2b. May already be entered by the Permit-issuing authority.	

OPERATING PROCEDURES	STEP SEQUENCE	INFORMATION/OPERATING GOALS/SPECIFICATIONS	TRAINING GUIDE NOTES
G. Suspended Solids, Final Effluent. (cont.)			

This portion of the Report Form, completed in accordance with assumed permit conditions, and the data of Table I, is shown in Fig. 6 below:

(32-37) PARAMETER		QUANTITY (3 card only) (38-45) MINIMUM	(46-53) AVERAGE	(54-61) MAXIMUM	UNITS	(62-63) NO. EX	CONCENTRATION (4 card only) (38-45) MINIMUM	(46-53) AVERAGE	(54-61) MAXIMUM	UNITS	(62-63) NO. EX	(64-68) FREQUENCY OF ANALYSIS	(69-70) SAMPLE TYPE
SUSPENDED SOLIDS	REPORTED	13.8	47.1	161	KG/DAY	–	12	25.0	60	MG/L	–	2/7	24-Hr. Comp
	PERMIT CONDITION	******	80	–			******	30	–			2/7	24-Hr. Comp

Fig. 6

H. Percent Removal Suspended Solids.			
1. Computation.	1. Calculate the percent removal of suspended solids for each pair of influent and effluent analyses made during the reporting period.		II.H.1 (p.33)
2. "Quantity" section	1. On the "Reported" line enter the minimum, average, and maximum percent removals in the corresponding spaces.	1a. Minimum - 66.6% 1b. Average - 86.5% 1c. Maximum - 93.0%	

EFFLUENT MONITORING PROCEDURE: REPORTING OF SELF-MONITORING DATA

OPERATING PROCEDURES	STEP SEQUENCE	INFORMATION/OPERATING GOALS/SPECIFICATIONS	TRAINING GUIDE NOTES
H. Percent Removal Suspended Solids.(cont.)	2. On the "Permit Condition" line enter the minimun and average removals required by the permit.	2a. There is no minimum removal requirement in the assumed permit conditions (Table II). Enter a dash in this space. 2b. Average removal required over a 30-consecutive-day period is 85% (Table II). 2c. May already be entered by the permit-issuing authority.	
	3. In the "No. Ex" column, on the "Reported" line, enter the number of times that the minimum percent removal required in the permit was not obtained.	3a. There is no minimum removal requirement in the assumed permit conditions (Table II). Enter a dash in this space. 3b. If none, enter a "0".	
3. "Concentration" section.	1. Enter dashes in the "Units" space and in the "No. Ex" space on the "Reported" line.	1a. May already be entered by the Permit-issuing authority.	

This portion of the form, completed in accordance with assumed permit conditions, and the data of Table I, is shown in Fig. 7 below:

PARAMETER		QUANTITY MINIMUM (30-45)	AVERAGE (46-53)	MAXIMUM (54-61)	UNITS	NO. EX (62-63)	CONCENTRATION MINIMUM (38-45)	AVERAGE (46-53)	MAXIMUM (54-61)	UNITS	NO. EX (62-63)	FREQUENCY OF ANALYSIS (64-68)	SAMPLE TYPE (69-70)
PERCENT REMOVAL SUSPENDED SOLIDS	REPORTED	66.6	86.5	93.0	%	-	*****	*****	*****		-	*****	*****
	PERMIT CONDITION	-	85	*****			*****	*****	*****	-		*****	*****

Fig. 7

EFFLUENT MONITORING PROCEDURE: REPORTING OF SELF-MONITORING DATA

OPERATING PROCEDURES	STEP SEQUENCE	INFORMATION/OPERATING GOALS/SPECIFICATIONS	TRAINING GUIDE NOTES
I. Fecal Coliform			
1. Computation	1. Calculate the geometric mean for the fecal coliform data obtained during the reporting period.	1a. See also the effluent monitoring procedure Calculation of the Geometric Mean of Coliform Counts by the Use of Logarithms.	II.I.1 (p. 34)
2. "Quantity" section	1. Place dashes in the "Units" space and in the "No. Ex" space on the "Reported" line.	1a. May already be entered by the Permit-issuing authority.	
3. "Concentration" section.	1. Enter the minimum and maximum reported results in the corresponding spaces on the "Reported" line.	1a. Minimum - 110 organisms/100 ml 1b. Maximum - 540 organisms/100 ml	
	2. Enter the geometric mean in the "average" space on the "Reported" line.	2a. Geometric Mean - 220 organisms/100 ml	
	3. On the "Permit Condition" line enter the geometric mean and maximum count specified in the permit.	3a. The geometric mean of samples analyzed over a 30-consecutive-day period is not to exceed 200 organisms/100 ml. (Table II) Enter "200" in "average" space. 3b. There is no maximum value specified in the permit (Table II). Enter a dash in the "maximum" space.	

EFFLUENT MONITORING PROCEDURE: REPORTING OF SELF-MONITORING DATA

OPERATING PROCEDURES	STEP SEQUENCE	INFORMATION/OPERATING GOALS/SPECIFICATIONS	TRAINING GUIDE NOTES
I. Fecal Coliform. (cont.)	4. In the "No. Ex" column, on the "Reported" line, enter the number of times during the reporting period that the maximum count allowed in the permit has been exceeded.	4a. If none, enter a "0". 4b. Since no maximum count is specified in the permit (Table II) enter a dash in this space.	
4. "Frequency of Analysis" column	1. Enter frequency of analysis during the reporting period on the "Reported" line.	1a. Enter 2/7, indicating that the analysis was performed twice weekly (Table I).	
	2. Enter required frequency of analysis, as specified in permit, on the "Permit Condition" line.	2a. Permit requires analysis twice weekly (Table II). Enter 2/7.	
5. "Sample Type" column.	1. Enter type of sample on which the analysis was performed.	1a. Enter "Grab", since it is assumed in this procedure that the sample has been obtained in accordance with permit requirements.	

This portion of the form, completed in accordance with assumed permit conditions, and the data of Table I, is shown in Fig. 8 below:

PARAMETER		(3 card only) (36-45) QUANTITY MINIMUM	(46-53) AVERAGE	(54-61) MAXIMUM	UNITS	(62-63) NO. EX	(4 card only) (38-45) CONCENTRATION MINIMUM	(46-53) AVERAGE	(54-61) MAXIMUM	UNITS	(62-63) NO. EX	(64-68) FREQUENCY OF ANALYSIS	(69-70) SAMPLE TYPE
FECAL COLIFORM	REPORTED	******	******	******	-	-	110	220	540	N/100ML	-	2/7	Grab
	PERMIT CONDITION	******	******	******			******	200	-			2/7	GRAB

Fig. 8

OPERATING PROCEDURES	STEP SEQUENCE	INFORMATION/OPERATING GOALS/SPECIFICATIONS	TRAINING GUIDE NOTES
J. Signature	1. The completed form must be signed by the ranking elected official of the municipality, or other duly authorized municipal employee.		
	2. Complete the four spaces provided at the bottom of the form.	2a. Name, title and signature of ranking elected official or duly authorized employee, and date of completion.	
	3. Forward completed form to permit-issuing authority in accordance with reporting instructions specified in permit.	3a. The entire form, completed in accordance with the permit conditions assumed for the purpose of this procedure, is shown in Fig. 9.	

NATIONAL POLLUTANT DISCHARGE ELIMINATION SYSTEM
DISCHARGE MONITORING REPORT

Form Approved
OMB NO. 158-R0073

City of Noname, Dept. of Environmental Services
184 Any Street
Noname, Anystate, 12345

INSTRUCTIONS

1. Provide dates for period covered by this report in spaces marked "REPORTING PERIOD".
2. Enter reported minimum, average and maximum values under "QUANTITY" and "CONCENTRATION" in the units specified for each parameter as appropriate. Do not enter values in boxes containing asterisks. "AVERAGE" is average computed over actual time discharge is operating. "MAXIMUM" and "MINIMUM" are extreme values observed during the reporting period.
3. Specify the number of analyzed samples that exceed the maximum (and/or minimum as appropriate) permit conditions in the columns labeled "No. Ex." If none, enter "0".
4. Specify frequency of analysis for each parameter as No. analyses/No. days. (e.g., "3/7" is equivalent to 3 analyses performed every 7 days.) If continuous enter "CONT."
5. Specify sample type ("grab" or "___ hr. composite") as applicable. If frequency was continuous, enter "NA".
6. Appropriate signature is required on bottom of this form.
7. Remove carbon and retain copy for your records.
8. Fold along dotted lines, staple and mail Original to office specified in permit.

		PERMIT NUMBER	DIS	SIC	LATITUDE	LONGITUDE
AN ST		N01234567	001	4952	47°20'4"	98°26'11"

REPORTING PERIOD: FROM 7|4 0|9 0|1 (YEAR MO DAY) TO 7|4 0|9 3|0 (YEAR MO DAY)

PARAMETER		QUANTITY MINIMUM	QUANTITY AVERAGE	QUANTITY MAXIMUM	UNITS	NO. EX	CONCENTRATION MINIMUM	CONCENTRATION AVERAGE	CONCENTRATION MAXIMUM	UNITS	NO. EX	FREQUENCY OF ANALYSIS	SAMPLE TYPE
FLOW	REPORTED	0.15	0.49	0.81	MGD	-	******	******	******		-	Cont.	-
	PERMIT CONDITION	******	-	-			******	******	******	-		Cont.	-
PH	REPORTED	5.4	******	9.2	STANDARD UNITS	3	******	******	******		-	7/7	Grab
	PERMIT CONDITION	6.0	******	9.0			******	******	******	-		2/7	Grab
BOD 5	REPORTED	11.0	37.5	94	KG/DAY	-	15	21	35	MG/L	-	2/7	24-Hr.Comp
	PERMIT CONDITION	******	70	-			******	30	-			2/7	24-Hr.Comp
PERCENT REMOVAL BOD 5	REPORTED	77	87.9	91	%	-	******	******	******		-	******	******
	PERMIT CONDITION	-	85	******			******	******	******	-		******	******
SUSPENDED SOLIDS	REPORTED	13.8	47.1	161	KG/DAY	-	12	25.0	60	MG/L	-	2/7	24-Hr.Comp
	PERMIT CONDITION	******	80	-			******	30	-			2/7	24-Hr.Comp
PERCENT REMOVAL SUSPENDED SOLIDS	REPORTED	66.6	86.5	93.0	%	-	******	******	******		-	******	******
	PERMIT CONDITION	-	85	******			******	******	******	-		******	******
FECAL COLIFORM	REPORTED	******	******	******	-	-	110	220	540	N/100ML	-	2/7	Grab
	PERMIT CONDITION	******	******	******			******	200	-			2/7	GRAB
	REPORTED												
	PERMIT CONDITION												

NAME OF PRINCIPAL EXECUTIVE OFFICER	TITLE OF THE OFFICER	DATE		
Doe (LAST) John (FIRST) J. (MI)	Mayor (TITLE)	7|4 0|9 3|0 (YEAR MO DAY)	I certify that I am familiar with the information contained in this report and that to the best of my knowledge and belief such information is true, complete, and accurate.	John J. Doe — SIGNATURE OF PRINCIPAL EXECUTIVE OFFICER OR AUTHORIZED AGENT

T-40 (4-74)

PAGE 1 OF 1

Fig. 9

EFFLUENT MONITORING PROCEDURE: REPORTING OF SELF-MONITORING DATA

<u>TRAINING GUIDE</u>

<u>SECTION</u>	<u>TOPIC</u>
*I	Introduction
*II	Educational Concepts - Mathematics
III	Educational Concepts - Science
IV	Educational Concepts - Communication
V	Field & Laboratory Equipment
VI	Field & Laboratory Reagents
VII	Field & Laboratory Analysis
VIII	Safety
*IX	Records & Reports

Training guide materials are presented here under the headings marked.
These standardized headings are used throughout this series of procedures

INTRODUCTION	Section I
TRAINING GUIDE NOTE	REFERENCES/RESOURCES

A	Holders of discharge permits issued by the U. S. Environmental Protection Agency are to report self-monitoring data either on EPA Form T-40 (Fig. 10), or on EPA Form 3320-1 (Fig. 11). Form T-40 will be used temporarily, and will be furnished by EPA to municipalities for reporting purposes. The T-40 form shown in Fig. 10 consists of the 3320-1 form, which has been overprinted for the reporting of data for basic parameters common to all municipal wastewater discharges. As information for each municipality is incorporated into EPA's computer system, form 3320-1 will replace Form T-40. For data reporting, a municipality will then receive from EPA Form 3320-1 on which the effluent parameters specific to that municipality will be computer overprinted. Until that time, however, data for any additional parameters to be reported which are not now included in the overprint on T-40 will be entered by the municipality, using as many additional blank copies of the form as are required. Completion of form T-40, is illustrated in this procedure for the basic parameters, assuming that permit conditions are as indicated in Table II. Reporting of additional parameters would be done in a manner similar to that illustrated.	

NATIONAL POLLUTANT DISCHARGE ELIMINATION SYSTEM

DISCHARGE MONITORING REPORT

Form Approved
OMB NO. 158-R0073

(2-3) ST	(4-16) PERMIT NUMBER	(17-19) DIS	SIC	LATITUDE	LONGITUDE

REPORTING PERIOD: FROM (20-21) YEAR (22-23) MO (24-25) DAY TO (26-27) YEAR (28-29) MO (30-31) DAY

INSTRUCTIONS

1. Provide dates for period covered by this report in spaces marked "REPORTING PERIOD".
2. Enter reported minimum, average and maximum values under "QUANTITY" and "CONCENTRATION" in the units specified for each parameter as appropriate. Do not enter values in boxes containing asterisks. "AVERAGE" is average computed over actual time discharge is operating. "MAXIMUM" and "MINIMUM" are extreme values observed during the reporting period.
3. Specify the number of analyzed samples that exceed the maximum *(and/or minimum as appropriate)* permit conditions in the columns labeled "No. Ex." If none, enter "O".
4. Specify frequency of analysis for each parameter as No. analyses/No. days. *(e.g., "3/7" is equivalent to 3 analyses performed every 7 days.)* If continuous enter "CONT."
5. Specify sample type *("grab" or "__ hr. composite")* as applicable. If frequency was continuous, enter "NA".
6. Appropriate signature is required on bottom of this form.
7. Remove carbon and retain copy for your records.
8. Fold along dotted lines, staple and mail Original to office specified in permit.

(32-37) PARAMETER		QUANTITY (3 card only) MINIMUM (38-45)	QUANTITY AVERAGE (46-53)	QUANTITY MAXIMUM (54-61)	UNITS	(62-63) NO. EX	CONCENTRATION (4 card only) MINIMUM (38-45)	CONCENTRATION AVERAGE (46-53)	CONCENTRATION MAXIMUM (54-61)	UNITS	(62-63) NO. EX	(64-68) FREQUENCY OF ANALYSIS	(69-70) SAMPLE TYPE
FLOW	REPORTED				MGD		******	******	******				
	PERMIT CONDITION	******					******	******	******				
PH	REPORTED		******		STANDARD UNITS		******	******	******				
	PERMIT CONDITION		******				******	******	******				
BOD 5	REPORTED				KG/DAY					MG/L			
	PERMIT CONDITION	******					******						
PERCENT REMOVAL BOD 5	REPORTED				%		******	******	******			******	******
	PERMIT CONDITION			******			******	******	******			******	******
SUSPENDED SOLIDS	REPORTED				KG/DAY					MG/L			
	PERMIT CONDITION	******					******						
PERCENT REMOVAL SUSPENDED SOLIDS	REPORTED				%		******	******	******			******	******
	PERMIT CONDITION			******			******	******	******			******	******
FECAL COLIFORM	REPORTED	******	******	******						N/100ML			
	PERMIT CONDITION	******	******	******			******						GRAB
	REPORTED												
	PERMIT CONDITION												

NAME OF PRINCIPAL EXECUTIVE OFFICER	TITLE OF THE OFFICER	DATE	*I certify that I am familiar with the information contained in this report and that to the best of my knowledge and belief such information is true, complete, and accurate.*	
LAST FIRST MI	TITLE	YEAR MO DAY		SIGNATURE OF PRINCIPAL EXECUTIVE OFFICER OR AUTHORIZED AGENT

T-40 (4-74)

PAGE OF

Fig. 10

NATIONAL POLLUTANT DISCHARGE ELIMINATION SYSTEM
DISCHARGE MONITORING REPORT

Form Approved
OMB NO. 158-R0073

ST | PERMIT NUMBER (4-16) | DIS (17-19) | SIC | LATITUDE | LONGITUDE

REPORTING PERIOD: FROM (20-21) (22-23) (24-25) YEAR MO DAY   TO (26-27) (28-29) (30-31) YEAR MO DAY

INSTRUCTIONS

1. Provide dates for period covered by this report in spaces marked "REPORTING PERIOD".
2. Enter reported minimum, average and maximum values under "QUANTITY" and "CONCENTRATION" in the units specified for each parameter as appropriate. Do not enter values in boxes containing asterisks. "AVERAGE" is average computed over actual time discharge is operating. "MAXIMUM" and "MINIMUM" are extreme values observed during the reporting period.
3. Specify the number of analyzed samples that exceed the maximum (and/or minimum as appropriate) permit conditions in the columns labeled "No. Ex." If none, enter "O".
4. Specify frequency of analysis for each parameter as No. analyses/No. days. (e.g., "3/7" is equivalent to 3 analyses performed every 7 days.) If continuous enter "CONT."
5. Specify sample type ("grab" or "___ hr. composite") as applicable. If frequency was continuous, enter "NA".
6. Appropriate signature is required on bottom of this form.
7. Remove carbon and retain copy for your records.
8. Fold along dotted lines, staple and mail Original to office specified in permit.

| PARAMETER (32-37) | | QUANTITY (3 card only) | | | | | CONCENTRATION (4 card only) | | | | | FREQUENCY OF ANALYSIS (64-68) | SAMPLE TYPE (69-70) |
		MINIMUM (38-45)	AVERAGE (46-53)	MAXIMUM (54-61)	UNITS	NO. EX (62-63)	MINIMUM (38-45)	AVERAGE (46-53)	MAXIMUM (54-61)	UNITS	NO. EX (62-63)		
	REPORTED												
	PERMIT CONDITION												
	REPORTED												
	PERMIT CONDITION												
	REPORTED												
	PERMIT CONDITION												
	REPORTED												
	PERMIT CONDITION												
	REPORTED												
	PERMIT CONDITION												
	REPORTED												
	PERMIT CONDITION												
	REPORTED												
	PERMIT CONDITION												
	REPORTED												
	PERMIT CONDITION												

NAME OF PRINCIPAL EXECUTIVE OFFICER	TITLE OF THE OFFICER	DATE	
		YEAR MO DAY	I certify that I am familiar with the information contained in this report and that to the best of my knowledge and belief such information is true, complete, and accurate.
LAST FIRST MI	TITLE		SIGNATURE OF PRINCIPAL EXECUTIVE OFFICER OR AUTHORIZED AGENT

EPA Form 3320-1 (10-72)

PAGE ___ OF ___

Fig. 11

EDUCATIONAL CONCEPTS - MATHEMATICS	Section II
TRAINING GUIDE NOTE	REFERENCES/RESOURCES

	NOTE: In all of the calculations in this section, rules for computation as given by Crumpler and Yoe were followed. Crumpler, T. B. and Yoe, J. H. <u>Chemical Computations and Errors.</u> Wiley and Sons. N.Y. 1940.	Crumpler, T. B. and Yoe, J. H. <u>Chemical Computations and Errors.</u> Wiley and Sons N.Y. 1940.
E.1	Computation of Quantities of BOD discharged in Final Effluent.	

Pertinent reported data from Table I is listed in the first three columns of the Table below. In column 4 the flow has been converted from gpd to MGD. The quantity of BOD_5 is obtained by multiplying the value in column 3 (mg/l) by the value in column 4 (MGD), and then multiplying the result by the factor 3.78.

This is expressed in mathematical form as

Kg/day = MGD x mg/l x 3.78

Example:

On September 3, Kg/day = 0.33 x 16 x 3.78 = 20

Date	Flow gpd	BOD_5 mg/l	Flow MGD	BOD_5 Kg/day
3	326,900	16	0.33	20
6	458,500	15	0.46	26
9	146,000	20	0.146	11.0
12	519,200	20	0.52	39
16	708,900	35	0.71	94
20	272,900	19	0.27	19
23	761,800	20	0.76	57
26	291,600	20	0.29	22
29	525,600	25	0.53	50
Total		190		338

$$\text{Average } BOD_5 = \frac{190}{9} = 21 \text{ mg/l}$$

$$\text{Average } BOD_5 = \frac{338}{9} = 37.5 \text{ Kg/day}$$

EDUCATIONAL CONCEPTS - MATHEMATICS	Section II
TRAINING GUIDE NOTE	REFERENCES/RESOURCES

F.1

Computation of percent BOD_5 removals.

Pertinent reported data from Table I is listed in the first three columns of the Table below. The percent BOD_5 removal for each day appears in column 4.

The percent removal is obtained by subtracting the concentration of BOD_5 in the final effluent from that in the plant influent, dividing this difference by the concentration of BOD_5 in the influent, and multiplying the result by 100. This can be expressed in mathematical form as follows:

$$\%BOD_5 \text{ Removal} = \frac{\text{Influent } BOD_5 \text{ (mg/1) } - \text{ Effluent } BOD_5 \text{ (mg/1)}}{\text{Influent } BOD_5 \text{ (mg/1)}} \times 100$$

Example:

On September 3, % BOD Removal $= \dfrac{170 - 16}{170} \times 100 = 91\%$

Date	BOD_5-mg/1 Inf.	Eff.	% Removal
3	170	16	91
6	160	15	91
9	200	20	90
12	190	20	89
16	150	35	77
20	170	19	89
23	150	20	87
26	190	20	89
29	190	25	87
Total	1,570	190	
Average	174	21	

Average BOD_5 Removal $= \dfrac{174-21}{174} \times 100 = 87.9\%$

EDUCATIONAL CONCEPTS - MATHEMATICS	Section II
TRAINING GUIDE NOTE	REFERENCES/RESOURCES

G.1

Computation of Quantities of Suspended Solids Discharged in Final Effluent.

Pertinent reported data from Table I is listed in the first three columns of the Table below. In column 4 the flow has been converted from gpd to MGD. The quantity of suspended solids is obtained by multiplying the value in column 3 (mg/l) by the value in column 4 (MGD), and then multiplying the result by the factor 3.78. This is expressed in mathematical form as:

$$KG/day = MGD \times mg/l \times 3.78$$

Example: On September 3, Kg/day = 0.33 x 12 x 3.78 = 15

Date	Flow gpd	T.S.S. mg/l	Flow MGD	T.S.S. Kg/day
3	326,900	12	0.33	15
6	458,500	16	0.46	28
9	146,000	25	0.146	13.8
12	519,200	25	0.52	49
16	708,900	60	0.71	161
20	272,900	19	0.27	19
23	761,800	23	0.76	66
26	291,600	20	0.29	22
29	525,600	25	0.53	50
Total		225		424

$$\text{Average T.S.S.} = \frac{225}{9} = 25.0 \text{ mg/l}$$

$$\text{Average T.S.S.} = \frac{424}{9} = 47.1 \text{ kg/day}$$

EDUCATIONAL CONCEPTS - MATHEMATICS	Section II
TRAINING GUIDE NOTE	REFERENCES/RESOURCES

H.1

Computation of percent suspended solids removal.

Pertinent reported data from Table I is listed in the first three columns of the Table below. The percent suspended solids removal for each day appears in column 4.

The percent removal is obtained by subtracting the concentration of suspended solids in the final effluent from that in the plant influent, dividing this difference by the concentration of suspended solids in the plant influent, and multiplying the result by 100. This can be expressed in mathematical form as follows:

$$\% \text{ T.S.S. Removal} = \frac{\text{Influent T.S.S. (mg/l) - effluent T.S.S. (mg/l)}}{\text{Influent T.S.S. (mg/l)}} \times 100$$

Example:

On September 3, $\% \text{ T.S.S. removal} = \frac{171 - 12}{171} \times 100 = 93.0\%$

Date	T.S.S.-mg/l Inf.	Eff.	% Removal
3	171	12	93.0
6	168	16	90.5
9	200	25	87.5
12	198	25	87.4
16	180	60	66.6
20	170	19	88.8
23	186	23	87.6
26	195	20	89.7
29	195	25	87.2
Total	1,663	225	
Average	185	25	

$$\text{Average \% Removal} = \frac{185-25}{185} \times 100 = 86.5\%$$

EDUCATIONAL CONCEPTS - MATHEMATICS	Section II
TRAINING GUIDE NOTE	REFERENCES/RESOURCES

I.1

Computation of the Geometric Mean

Pertinent reported data from Table I is listed in the first two columns of the Table below. The logarithm of the reported Coliform value appears in column 3. Note that two-place logarithms are used in this calculation. That is, the mantissa of the logarithm (the numbers to the right of the decimal point) contains only two numbers. Two-place logarithms are adequate since the coliform values are reported only to two significant figures.

The logarithms in column 3 are added, the total is divided by the number of values reported, and the anti-logarithm of the quotient is obtained. This is the geometric mean. It is reported to two significant figures.

Date	Fecal Coliform N/100 ml	Log of Fecal Coliform
3	350	2.54
6	540	2.73
9	180	2.26
12	170	2.23
16	220	2.34
20	240	2.38
23	110	2.04
26	130	2.11
29	280	2.45
	Total	21.08

$$\frac{21.08}{9} = 2.34$$

The antilogarithm of 2.34 is 220. This is the geometric mean. If the antilogarithm did not end with a zero, the number would be rounded to the nearest ten for reporting purposes.

RECORDS AND REPORTS	Section IX
TRAINING GUIDE NOTE	REFERENCES/RESOURCES

C.1.1.

Reporting of minimum and/or maximum values may or may not be required by the permit-issuing authority. If not required, a dash or an asterisk may already be entered in either or both of these spaces. The same is true for all of the other parameters shown, with the exception of pH, for which minimum and maximum values must be reported.

C.1.3.3a

Overprinted forms may already have either a dash or an asterisk in this space. This also applies to all other cases in this procedure where the entry of a dash in a space is specified.